BEYOND THE APES & VAPES OF WEB3

Why you should be part of the Web3 journey

Renee Francis

A catalogue record for this book is available from the National Library of Australia

NATIONAL LIBRARY OF AUSTRALIA

This book is non-fiction.

Publisher:
Inspiring Publishers
P.O. Box 159, Calwell, ACT Australia 2905
Email: inspiringpublisher.com
http://www.inspiringpublishers.com

National Library of Australia Cataloguing-in-Publication entry

Author: Renee Francis

Title: **Beyond the Apes and Vapes of Web3**

ISBN: 978-1-923087-56-9 (Print)
ISBN: 978-1-923087-55-2 (ePub2)

Introduction:
A Journey of Discovery

Okay, I'll spill the beans. Writing this book wasn't really on my to-do list. I'm the kind of person who values privacy, but the idea of putting my thoughts and experiences into a book wasn't so bad. Then came the dread of having to talk about it and actually give it out to people. The mere thought made me cringe. Can you relate?

It was only about a year ago that I mustered up the courage to start sharing updates about my business on LinkedIn. I never thought I'd be someone who'd put myself out there like that. But you know what? Something unexpected happened. I started gaining a following. People actually found value in what I had to say. It was a pleasant surprise.

For years, friends and colleagues had been telling me to share my knowledge and experiences. They believed I had something helpful to offer, even though I wasn't convinced. A year later I realised they were onto something. I found myself in a position where I had insights and experiences that could actually make a difference for others going through similar journeys in life and business.

And you know what? Helping people is something I genuinely want to do. So, when my mentor suggested I write a book, my initial reaction was to wince and resist. But after giving it some thought, I realised that this was just another way to reach and help even more people. So, here we are, with me putting pen to paper (or fingers to keyboard) and sharing my knowledge, experiences, and insights for the benefit of others.

It's funny how life works sometimes, isn't it? We never quite know where our journey will take us. But if there's an opportunity to make a positive impact, it's worth pushing past our comfort zones and embracing it. And that's exactly what I'm doing with this book.

Together, we'll dive into the lessons, the wins, and the failures that have shaped my path into the Web3 world. Let's embark on this adventure of learning and growth together. I'm excited to offer whatever insights I can to help you along your own journey.

I started my career in a typical way - landing a corporate job right after (or even during) university. I settled into a stable role at one company for a couple of years, but then, over the course of 10 years, I found myself working in 10 different companies. Talk about variety!

Those years taught me so much about business, teams, different industries, and management styles. I learned how to navigate projects with abundant resources and, at times, with limited resources too. It was a ride of gaining valuable experience and getting results in different offices with different styles. But I had this nagging feeling that something wasn't quite right. And it wasn't just me - I talked to so many others who felt the same way. It was that centralised structure that rubbed us the wrong way. It felt suffocating, like there had to be a better way.

One day, I realised that if I kept going down this path, I would be doing the same thing for the next five decades. Looking around, I saw colleagues who had been with the same company for 5, 10, or even 20 years, and I knew that wasn't the life I wanted for myself.

So, I made a decision. I needed to find another way, a path that aligned with my aspirations and desires. Life is too short and too precious to settle for something that doesn't fulfill us. And that's why I ventured out to seek a different kind of journey, one that would allow me to chart my own path, explore my passions, and make a meaningful impact on the world. I was seeking a life with more flexibility and freedom.

The corporate world provided a solid foundation and I learnt so much, but it was time for me to find a path that would truly light me up. And you know what? I haven't looked back. The journey may not always be straightforward, but the freedom and fulfillment I've found are worth every twist and turn.

Those big global corporations seemed to be all about centralised structures, rigidity, and control. Their one-size-fits-all approach didn't fit me. I'm all about creativity, thinking outside the box, and coming up with fresh ideas. I don't believe in doing the same thing over and over again. And as for societal norms and power dynamics – they just don't sit well with me. I'm all about breaking free from those constraints and finding a better way.

Back in 2016, I started my own digital marketing agency called The Bubble Co. And let me tell you, I had a crystal-clear vision right from the get-go. I wanted to create a work environment that was remote and flexible, breaking free from the old-school norms. I wanted a place where everyone's contributions mattered, where people were treated as human beings rather than mere numbers on a spreadsheet.

When you think about it, our working structures were born during the Industrial Revolution, over 200 years ago. So much has changed since then, but our way of working had remained stuck in the past. It was high time for a shake-up. I was determined to be a part of that change.

I believed that hard work and dedication should be rewarded with trust and flexibility. After all, why should we be confined to outdated 9-to-5 schedules and restrictive office spaces? The world was evolving, and so should our approach to work. I wanted to create an environment where people could thrive, where their ideas could flourish, and where they could achieve a healthy work-life balance.

So, I set out to drive this better way of working. It wasn't always easy, and there were certainly challenges along the way. But seeing the positive impact it had on my team, the sense of empowerment and fulfillment they experience, makes it all worth it.

Now, here we are, living in a world where remote work and flexible arrangements have become the norm for many. It's incredible to witness how much has changed in such a short time. And I'm proud to have played my part in driving that change. I will keep pushing for work environments that value the individual, that prioritise flexibility, and that foster a culture of trust and respect. Because when we break free from the old structures and embrace a new way of working, we unlock incredible potential and create a better future for all.

I made a huge discovery back in 2019/2020 - Web3, blockchain, crypto and decentralisation. It was like a lightbulb moment, a realisation that spoke directly to my soul. Suddenly, everything made sense, and I saw a way to break free from those rigid structures that had been holding us back. This was our ticket to a more decentralised and open future, and I knew I had found my calling.

I'm on a mission to empower others to join me in shaping the future of the decentralised web. I want to share my learnings and experiences in Web3, to spread the word and gain even more momentum in this revolutionary movement. It's all about creating a space where power is distributed, where everyone has a voice, and where we can build a future that truly represents our collective values.

I'll share how I discovered Web3, the moment it clicked for me, and why it resonates with me so deeply. It's not just about the technology itself - it's about the potential it holds, the opportunities it brings, and the power it gives back to the people.

Together, we can pave the way for a future where control isn't concentrated in the hands of a few, but is shared among the many. We can redefine the rules, rewrite the narrative, and create a more equitable and empowering digital landscape.

We'll explore its possibilities, unravel its complexities, and unlock its potential. This is our chance to shape the future, and I can't wait to have you by my side on this exciting journey. Get ready, because we're about to embark on something truly transformative.

PART 1:
THE AWAKENING

Chapter 1:
What is Web3?

To start describing Web3, we need to take a little journey back in time starting with Web1.

Web1 feels like forever ago, but it was the early days of the internet when things were pretty static. Picture this: dial-up internet, that distinctive 'dial-up internet sound' echoing through the house as we anxiously waited for a successful connection. Ah, the good old days. But here's the kicker - back then, if someone made a phone call, it would cut off our precious internet connection. Can you believe it? There was a time when we couldn't surf the web and talk on the phone simultaneously. Younger readers, this might sound like ancient history to you, but for us seasoned internet users, it was a reality.

So, in this Web1 era, the internet was mostly a 'read-only' experience. We could send emails, but there wasn't much else in terms of interactivity. No social media, no streaming videos, just the thrill of receiving an email in our inbox. It's fascinating to think about how far we've come since those simpler times.

As I was writing this book, I couldn't help but indulge in a bit of nostalgia. I actually Googled the 'dial-up internet sound' and let me tell you, the memories came flooding back. It's amazing how a sound can transport us back in time and remind us of where it all began. Web1 laid the groundwork for what was to come, and it's incredible to see how much the internet has evolved since then.

Let's talk about Web2, the internet that we're all familiar with as I write this book in 2023. This is the era of the internet where things got a lot more interactive and dynamic. Online banking, digital wallets like PayPal, social media platforms, cloud storage, and a whole range of features that make our lives easier and more connected are the norm. Web2 was a game-changer in terms of interactivity and participation. It brought us closer together and revolutionised the way we interact with technology.

But, and here's the big 'but', there's a downside to Web2. One of the major drawbacks is the centralisation of power. You see, the global tech giants have their fingerprints all over our data, apps, and content. They pretty much own and control everything. It's like they're the gatekeepers of the digital world, and that can be a bit unsettling. We're at their mercy, and it raises concerns about privacy, security, and the concentration of power in the hands of a few.

Don't get me wrong - Web2 has brought us incredible advance-ments and conveniences, and it's an integral part of our daily lives. But it's also important to acknowledge its limitations and the need for a more decentralised and inclusive approach. That's where Web3 comes in.

Web3 is all about shifting the power back to the people. It's about decentralisation, transparency, and giving individuals more control over their data and online experiences. With Web3, we have the potential to create a more equitable and user-centric digital landscape. It's like the next level of the internet, but with a twist. Instead of relying on big tech giants, Web3 is all about decentralisation and autonomy, thanks to the power of block-chain technology.

With Web3, users like you and me have more control and ownership over our online experiences. We're not just passive consumers anymore - we become active participants in this decentralised internet. Say goodbye to relying on those big tech giants for everything. In Web3, we get to enjoy increased security, have a say in how our data is used, protect our privacy, and best of all, we're free from the chains of censorship.

It's a promise of a better internet and a better future. Web3 gives us the opportunity to shape the digital landscape, to create a more inclusive and user-centric online world. It's all about empowering individuals, promoting transparency, and redefining how we interact with technology.

I'd like to add that there is a common misconception that Web3.0 and Web3 are the same. They are not. Software development company Leewayhertz has a clear definition of how they differ:

"Tim Berners-Lee coined the term Semantic Web, which refers to a version of the web that can connect everything at the data level. He stated that with the emergence of the semantic web, "the day-to-day mechanisms of trade, bureaucracy and our daily lives will be handled by machines talking to machines. The "intelligent agents" people have touted for ages will finally materialise."

In the present-day Internet, there are information silos. For instance, the information you upload on LinkedIn will not be automatically updated on Facebook or Twitter because they are not linked. Berners-Lee aimed to connect all the information by linking web pages and making them interoperable so that no one ever needed to upload their information separately on different online platforms.

From the discussion above, it's easy to infer that although people connect web 3.0 with Web3, they are not the same."

Back to Web3. Let's explore its main components. I'll break it down in simple terms so we can all wrap our heads around it:

First up is decentralisation. This is all about taking control and ownership away from those big tech giants and giving the power back to the people. No more monopolies or central authorities calling the shots. It's like a digital revolution where everyone gets a say.

Next is blockchain. This is like a decentralised public ledger that keeps track of transactions across multiple computers. It's not controlled by any single authority, which means it's more transparent and secure. Think of it as a collaborative record-keeping system that can't be tampered with easily.

Now, let's talk about cryptocurrency. It's like digital money that exists on the blockchain. Transactions using cryptocurrency are verified and stored on the blockchain, rather than relying on a centralised authority like a bank. It's a whole new way of handling transactions and exchanging value, without the need for intermediaries.

Lastly are smart contracts. These are like digital agreements between two parties, but instead of being written on paper, they are coded and executed on the blockchain. Smart contracts automatically enforce the terms of the agreement, eliminating the need for intermediaries or third parties. It's a secure and efficient way to make agreements and ensure everyone follows through.

Web3 is all about decentralisation, blockchain, cryptocurrency, and smart contracts. It's a new way of doing things that empowers individuals, promotes transparency, and creates a more efficient and inclusive digital ecosystem.

Below is a diagram that summarises the evolution of these web iterations.

Web 1.0
Read only era

Web 2.0
Read and write era

Web 3.0
Read, write and own era - we get to own a piece of it

Image 1. Source: thebubbleco.com.au

Chapter 2:
The Catalyst:
Why Web3 Matters to Me

My Web3 journey began with my introduction to cryptocurrency, specifically Bitcoin. It was around 2012/2013 when I first heard about it, although I didn't jump on the bandwagon back then. I remember sitting in my corporate office, listening to my IT colleagues talk about their investments in Bitcoin. One guy had put in $700, while the other had invested around $5,000. Little did we know that Bitcoin would skyrocket in value over the years, making those early investments quite lucrative.

Watching my colleagues stress over the price fluctuations was eye-opening. Cryptocurrency can be volatile, and the swings can happen within hours. Despite their anxieties, they encouraged me to get involved. But at that time, I wasn't convinced. Why would I want to get into something that caused so much stress? It didn't seem like a fun experience to me. One colleague eventually sold out, breaking even after being unable to handle the ride. The other made a tidy profit and treated himself to a new car. You can decide whether it was worth the daily stress for him.

In 2019 I attended an event headlined by author and coach Tony Robbins. It was a turning point for me as I had just left my full-time corporate job and was fully committed to running my own business. That event opened my eyes to topics like starting and scaling a business, the financial markets, flaws in our current monetary system, and personal development. I delved into more courses, devoured books by authors including Robert Kiyosaki, and immersed myself in the knowledge.

In 2020, as the world went into a pandemic-induced lockdown, I saw an opportunity to further my learning. After I received an email about a webinar featuring Robert Kiyosaki and other experts I eagerly signed up and absorbed all the knowledge I could. Surprisingly, it was one of the other speakers, Marcus De Maria from Investment Mastery, who caught my attention. His insights about the broken financial system and how blockchain and cryptocurrency were solving those problems resonated deeply with me. I was hooked. I recorded the webinar and imme-diately enrolled in his programs. It was a 'take my money' moment because I also believed in the potential of decentralisation and the promises of Web3.

From that point on, I was on a journey of investing and personal growth, fuelled by the idea of decentralisation. I was eager to see a more decentralised and equitable world. I couldn't help but think about all the parents I worked with during my corporate days who missed out on precious moments with their children due to the rigid structures of our work lives. The centralised control and systems that originated during the Industrial Revolution had hardly changed, despite the significant shifts in our lifestyles. I also realised that I had always encouraged those around me to take ownership of their lives and explore alternative paths.

I questioned why society pushed us towards a predefined life trajectory without teaching financial education to navigate a different route.

Over the next 12 months, these lessons changed the trajectory of my life and business. The concepts of decentralisation, blockchain, and Web3 reshaped my perspective and motivated me to integrate Web3 training and services into my digital marketing agency, The Bubble Co. We started educating our existing team members and gradually extended the knowledge to our Web2 clients. The more we learned and grew in the Web3 space, the more we realised its immense potential. This led me to launch a second agency, Take3, dedicated solely to Web3, where we hired Web3 natives to join our team. The growth and success we achieved in a short time have been truly remarkable, positioning us as one of the leading Web3 agencies in Australia by mid-2023.

As I continued my Web3 journey, I discovered that it encompassed much more than just cryptocurrency. Web3 represented a paradigm shift, a new way of approaching the internet and digital interactions. It was about decentralisation, autonomy, and empowering individuals in the digital realm. I realised that Web3 had the potential to reshape the foundations of our online world, enabling a more equitable and user-centric environment.

One of the fundamental concepts of Web3 is decentralisation. Unlike traditional Web2 platforms, which are controlled by central authorities or tech giants, Web3 aims to distribute power and decision-making among its users. It seeks to create a more democratic and inclusive digital landscape, where individuals have greater control over their data, privacy, and online experiences. By leveraging blockchain technology, Web3 removes the need for intermediaries and allows for direct peer-to-peer inter-actions, fostering trust and transparency.

Blockchain, a key component of Web3, serves as a decen-tralised public ledger that records transactions across multiple computers or nodes. It eliminates the need for a central authority to validate and authenticate transactions, as the blockchain's distributed network verifies and secures the data. This not only enhances security but also promotes transparency and immuta-bility, ensuring that transactions are tamper-proof and resistant to censorship.

Cryptocurrency is another Web3 aspect that captured my attention. It represents a digital form of currency that operates independently of traditional financial institutions. Transactions are verified and recorded on the blockchain, providing a trans-parent and secure method of exchanging value. Cryptocurrencies like Bitcoin and Ethereum have gained widespread recognition, challenging the existing financial system and offering alternative means of payment and investment.

Smart contracts emerged as a revolutionary concept within Web3. These self-executing contracts automatically enforce the terms and conditions defined within their code. Smart contracts eliminate the need for intermediaries or third parties in various scenarios, such as real estate transactions, supply chain manage-ment, or even crowdfunding campaigns. They enhance efficiency, reduce costs, and ensure that agreements are executed as intended, without relying on trust in a central authority.

Web3 not only presents technological advancements but also holds the potential to redefine social and economic structures. It opens up avenues for decentralised governance, where commu-nities can actively participate in decision-making processes.

DAOs (Decentralised Autonomous Organisations) enable collective decision-making and resource allocation through community voting and governance mechanisms, without the need for centralised entities. More on that later.

Furthermore, Web3 introduces the concept of digital identity, where individuals have control over their personal information and can selectively share it as needed. This empowers users to protect their privacy and mitigate risks associated with centralised data breaches or surveillance. Web3 envisions a future where individuals have sovereignty over their digital identities, fostering a more secure and trustworthy online environment.

The promise of individual ownership, reduced reliance on centralised entities like governments and big tech corporations, and the transformative power of decentralisation have convinced me that this is where the world is heading. I'm fully committed to playing my part in shaping that future, and I'm excited to see what lies ahead.

The adoption of Web3 is not without its challenges. One of the primary hurdles lies in educating and familiarising individuals with the concepts and technologies underlying Web3. The transition from Web2 to Web3 requires a shift in mindset and understanding. It demands individuals embrace decentralisation, take responsibility for their digital interactions, and navigate the complexities of managing digital assets and wallets. Additionally, scalability remains a concern as Web3 applications aim to handle increasing user demands without compromising efficiency or security.

To drive mainstream adoption of Web3, bridging the gap between early adopters and the broader audience is crucial. Education and user-friendly interfaces play a vital role in making

Web3 accessible and intuitive. Simplifying complex concepts, providing clear guidelines for securing digital assets, and developing seamless user experiences will be instrumental in attracting and retaining users.

My Web3 journey has been transformative. It has opened my eyes to the immense potential of decentralisation, blockchain, and cryptocurrency in reshaping our digital landscape. While challenges exist, the promise of Web3 is too significant to ignore. It is a call to action, an invitation to actively participate in shaping the future of the internet and redefining how we interact, transact, and govern in the digital realm. Web3 is the future. Web3 is inevitable. And I'm here for it.

PART 2:
THE ESSENCE OF WEB3

Chapter 3:
Unleashing the Power of Decentralisation

You may be wondering: why does decentralisation matter? To determine what we need to build, we must first grasp what we are up against. For centuries, we've had central authorities control various aspects of our lives – and it's all worked out OK, right?

Wrong. Let me provide you with a few examples of over-centralisation. First, let me be clear. I don't believe that everything will ever or should ever be decentralised. However, there are certain aspects of our lives that will benefit greatly when we implement a more decentralised approach.

Here are some obvious examples of over-centralisation.

The Soviet Union under Joseph Stalin: During Stalin's rule from the 1920s to 1953, the Soviet Union experienced a high degree of centralisation of power. Stalin's policies, including forced collectivisation of agriculture and the implementation of Five-Year Plans, led to widespread famine, political repression, and the death of millions of people. The centralisation of power in the hands of the Communist Party resulted in severe human rights abuses and economic mismanagement.

Economic Crisis in Russia: Following the collapse of the Soviet Union, Russia underwent a rapid transition from a centrally planned economy to a market-based system in the 1990s. However, the process of economic reform was marred by a high degree of centralisation and corruption. The concentration of power and resources in the hands of a few individuals led to widespread economic instability, inequality, and poverty, adversely affecting the lives of many Russians.

The 2008 global financial crisis: Originating in the United States with the collapse of the subprime mortgage market, its impacts quickly spread across the globe. The crisis led to a severe economic downturn, resulting in widespread job losses, bankruptcies, and foreclosures. People faced financial hardships as investments and retirement savings were wiped out, and the value of homes plummeted. Governments implemented austerity measures and bailout packages, leading to cuts in public spending, reduced social services, and increased inequality. The crisis also exposed the vulnerabilities of the financial system and eroded trust in institutions, causing long-lasting psychological and emotional distress for individuals and communities.

Let's take this very recent example of the 2008 financial crisis (GFC) and imagine what it could have looked like from a more decentralised perspective.

A decentralised approach to the financial system could have potentially prevented or reduced the impact of the GFC. A decentralised system could have spread risk, allowed for better regulation tailored to local conditions, encouraged responsible practices, fostered competition, and enhanced overall resilience. As we witnessed during that period, the very few elite and powerful controlled the many.

In a decentralised blockchain (like Bitcoin for example) no one has to know or trust anyone else. Each member of the network has a copy of the exact same data in the form of a distributed ledger. No one single person or group has dominant control, instead all the users collectively retain control. Bitcoin's blockchain, for example, is a digital database managed among the nodes of a peer-to peer network. The data in this blockchain is immutable, which means it cannot be altered or reversed. Bitcoin uses a proof-of-work algorithm to validate transactions and add them to the blockchain. No central point of authority or control, yet everybody can see the publicly available data, but they can't alter or corrupt it.

Decentralisation aims to redistribute power, decision-making and resources from a central authority to various entities. It empowers individuals and communities with greater control over their data and resources. It enhances accountability, transparency, and efficiency by reducing bureaucracy and enabling greater oversight. Overall, decentralisation unleashes the potential for more inclusive, democratic, and resilient systems that better serve the diverse needs and aspirations of people.

We've seen other negative effects of centralisation coming from social media and tech giants such as Meta.

Meta (formerly Facebook) created privacy concerns based on the amount of data they were collecting. This resulted in Facebook losing 1 million users in Europe in 2018, and 44% of Facebook users aged 18 to 29 deleted the app from their phone. In 2022, $80bn was wiped off the value of Meta.

Privacy Australia published an article in 2023 about the terrifying ways Facebook is using your data including mining user data with our authorisation, misusing data, collecting images for facial recognition and collecting non-user data.

Now let me be clear and fair. I own and run a digital marketing agency where we rely on these platforms to effectively target the right people with the right message at the right time. As a consumer myself, I much prefer sharing some data to receive content and ads that are relevant and specific to my interests. My issue is: who is profiting from the use of our personal data? In the Web2 world, global tech giants like Meta profit from this data.

Web3 promises a better way; where individuals will choose who they share data with and how much data they share. Better still - they will personally be rewarded or compensated for sharing such data. This is one of the most powerful reasons why I believe in Web3.

Chapter 4:
Empowering the Individual: Web3's Promise

When it comes to personal empowerment, it's all about taking the reins of your own life and making choices that align with your goals and desires. I've always been a strong believer in individuals pushing for what they truly want, regardless of societal expectations, parental pressure, or the demands of bosses. It's about making decisions that prioritise your own well-being and happiness while still being mindful of others around you. That's the essence of personal empowerment—having the courage and conviction to chart your own path and live life on your own terms. It's not always easy, but it's incredibly rewarding to know that you have the power to shape your own destiny.

I can completely relate to the expectations and traditions that come from growing up in a traditional ethnic family. The path of going to school, getting a stable job, and following the expected trajectory of life was ingrained in me from an early age. It's what my parents knew and believed would lead to a successful and fulfilling life. And while there's nothing inherently wrong with that path, it simply didn't align with what I envisioned for myself.

I realised that personal empowerment meant taking charge of my own destiny and making choices that resonated with my true desires and passions. It meant breaking away from the expectations and norms that had been imposed on me. I wanted to create a life that was uniquely my own, one that was filled with excitement, growth, and the pursuit of my dreams.

The journey towards personal empowerment isn't always easy. It involves overcoming the fear of disappointing others and stepping outside of your comfort zone. It requires embracing the unknown and taking risks. But let me tell you, the rewards are absolutely worth it. I read an awesome quote from Steven Bartlett about this that goes like this, "if taking care of yourself means letting someone down, then let someone down."

By embracing personal empowerment, I've been able to explore different paths, chase my passions, and create a life that aligns with my authentic self. It hasn't been a linear journey, and there have been ups and downs along the way. But each step I've taken towards personal empowerment has brought me closer to a sense of fulfillment and joy that I wouldn't have experienced otherwise.

If you're reading this book about Web3 and personal empowerment, chances are you're seeking a different path too. You're searching for ways to discover a life that empowers you on your own terms.

Personal empowerment is about finding the courage to challenge the status quo, to question the expectations placed upon us, and to create a life that reflects our true desires. It's about embracing our individuality and pursuing the things that truly light us up inside.

So, if the traditional path doesn't resonate with you, know that you have the power to forge your own path. Surround yourself with supportive individuals who believe in your dreams and aspirations. Seek out mentors and role models who have charted their own course and found success and happiness outside of societal norms.

Remember, personal empowerment is a journey. It's about constantly learning, growing, and adapting. It's about being open to new experiences, taking risks, and never settling for less than what you truly deserve. Trust in your own abilities, follow your intuition, and believe that you have the power to create a life that is uniquely yours.

Social expectations are defined as implicit rules that govern one's reactions and beliefs in a way that is deemed acceptable by society. There is a quote by Brian Tracy that I love, "The biggest enemy of your success is your comfort zone."

Let's take a look at some examples of some societal norms or parental expectations imposed on us:

- Get a stable job and work there for a long time
- Stay at a job for 10 years to accrue long service leave
- Save money and not take risks
- Investing is risky
- Trust that the government, banks, corporates and big tech giants have your best interests in mind
- To have a successful career, you must have a university degree.

So how does this relate to Web3?

Picture a world where you have complete control over your finances, identity, and assets without the interference or fees of middlemen. Let me explain with a simple example. Right now, when you want to send money to someone, you go through the hassle of going to the bank or doing it online. And guess what? The bank might even charge you a fee just to move your own money! Can you believe that? And that's not all - sometimes the transfer gets held up for days. Ever wondered why the bank needs to hold on to YOUR money and make you wait? It's frustrating, right? And if you're sending money to someone in another country the process can become even more time-consuming and costly. It's like the banks are making us jump through hoops and pay extra for a simple transaction. But hold on, there's a better way.

Enter Web3, which when combined with blockchain technology, means we can cut out the middlemen, like banks, and take control of our financial transactions. No more unnecessary fees or delays. Instead, we can transfer funds directly to the intended recipient, quickly and securely. Imagine the freedom and convenience of instant, low-cost money transfers, whether it's to your neighbour or someone on the other side of the globe. Web3 opens up a world of possibilities, making financial transactions simpler, faster, and more cost-effective. This is about breaking free from the confines of traditional banking and embracing a new era of financial empowerment.

In a Web3 world with cryptocurrency, this is done almost instantly and for very little fees (sometimes none at all). There is no bank, no intermediary, no one controlling your cash - no one but you. Just the power of the internet, blockchain and decentralisation.

Here's another example: saving money. We've all been raised with the belief that stashing our cash in a bank is the wise and responsible thing to do. I mean, I can still vividly remember being handed a children's account from the bank when I was in primary school and encouraged to save up all my pocket money. But hold up! Did you know that in 2021, Choice revealed that this program was nothing more than a sneaky marketing strategy targeting kids, without actually teaching them about financial literacy? Talk about a major letdown!

So, here's the million-dollar question: Is saving money in a bank really worth it in the long run, especially when we consider the constant printing and inflation of money year after year? Think about it for a moment. While it may feel safe and secure to have our money sitting in a bank account, the reality is that its value is gradually eroded over time due to inflation. It's like watching your hard-earned dollars slowly lose their purchasing power while the cost of goods and services continues to rise. It's a losing game.

With Web3 and the concept of decentralised finance (DeFi), we have the opportunity to explore alternative ways to save and grow our money. Imagine having the ability to invest directly in digital assets, participate in decentralised lending and borrowing, and earn passive income through various decentralised protocols. It's like taking control of your financial future and breaking free from the limitations of traditional banking.

By leveraging blockchain technology, Web3 allows us to be part of a decentralised financial ecosystem that operates transparently and without the need for intermediaries. We can explore innovative financial instruments like decentralised stablecoins, yield farming, and liquidity pools. These concepts might sound a bit technical at first, but they hold the potential to revolutionise the way we save and grow our wealth.

In 2023, coming out of the pandemic and years of printing out money, inflation is a big issue worldwide. Rising costs of living and prices of everyday items are hurting people. It started even earlier when in mid-2022 the price of a standard iceberg lettuce in Australia hit an all-time high of $AUD10. To put that in perspective, they normally retail for about $AUD3.50 each.

Why does this happen?

Let's talk about something that might make your head spin a little: quantitative easing. Have you ever come across this fancy term thrown around by banks and governments? Essentially, it's when a central bank, like the US Federal Reserve, decides to pump more money into the economy by printing more of it. They claim it's to give the economy a boost, but here's the catch: it often leads to some serious inflation problems.

You see, when more money is thrown into circulation, the value of the existing currency starts to dwindle. It's like diluting a drink with water - the more water you add, the weaker the flavour becomes. The same goes for money. When there's an influx of cash, its purchasing power takes a nosedive. Suddenly, the things you used to buy for a certain amount of money become more expensive. It's like a never-ending game of price hikes, and it can seriously mess with the economy.

Now, let's think about the impact this has on your money and savings. If the money you earn and diligently save is worth less and less over time, what's the point of squirreling it away? It's like you're saving up for something, only to find out that the value of your savings keeps shrinking. It's frustrating, to say the least.

That's where Web3 steps in with its promise of a better financial future. With the power of blockchain technology, Web3 offers a decentralised alternative that can help protect your money from the sneaky claws of inflation. By shifting away from traditional banking and embracing decentralised finance, you can explore new ways of saving and growing your wealth.

Imagine a world where you can invest directly in digital assets, participate in decentralised lending and borrowing, and even earn passive income through various decentralised protocols. It allows you to be in control of your financial future and say goodbye to the whims of banks and governments. It's about creating a system where your money retains its value and helps you achieve your goals.

So, the next time you hear the term 'quantitative easing' or notice the effects of inflation creeping into your daily life, remember that there's an alternative out there.

Traditional currency is also known as fiat currency, a term used to describe the main currencies around the world such as AUD, USD, GBP. The problem is these currencies have an unlimited supply as we keep printing more, and they are owned and controlled by a central authority.

Bitcoin fixed many of these issues. Bitcoin is decentralised so it's not owned or controlled by one central authority. Plus, it has a limited supply – only 21 million Bitcoin will ever be created, which creates scarcity. When something is scarce, its value increases and it is therefore known to be deflationary. Think of other things in life that are scarce, such as gold and time. Gold is scarce, therefore it is valuable. Our time on earth is scarce and therefore it is also valuable. This is how Bitcoin and many other cryptocurrencies solve the issues with our traditional fiat system.

And one final example: how it is used for marketing data ownership. I mentioned in Chapter 3 about the power and control big tech giants have over our data. You may have heard the saying from Netflix's 2020 documentary The Social Dilemma, "if you're not paying for the product, then you are the product." When we created our Facebook, Google, Instagram, LinkedIn and other social media accounts we were most likely drawn into the excitement of staying connected, learning new things, sharing the fun in our lives and following the lives of others we admired. Then the ads started appearing. Sometimes the ads were so specific or so timely, we even accused Facebook of listening to our conversations or 'stalking' us. This overreach and overuse of data is where the Web2 data ownership problem lies.

If you haven't heard of the Brave Browser, it's worth checking out. A Web3 browser that promises privacy and security, it is a seamless transition from Google Chrome, putting user's privacy first and allowing them to retain ownership over their data and how it is use for advertising purposes. "Browse privately. Search privately. And ditch Big Tech". That's Brave's slogan, vowing to do better than its Web2 counterparts.

There are several decentralised Web3 social platforms yet none have really gained mainstream adoption just yet. Let's look at two examples: Mastodon and Deso.

Mastodon is a decentralised, open-source social media platform that emerged as an alternative to traditional, centralised social media networks. It was created by Eugen Rochko and launched in 2016. Mastodon operates on a federated model, which means that instead of relying on a single central server, it consists of a network of independently operated servers called 'instances.'

Each Mastodon instance is like a separate community with its own rules and moderation policies, allowing users to join the instance that aligns with their interests or preferences. These instances are connected through a protocol called ActivityPub, which enables users from different instances to interact with each other.

Mastodon provides features similar to other social media platforms, including the ability to post text-based updates, share images, videos, and links, as well as engage in discussions through comments and direct messages. It also allows users to follow and be followed by others, create and join groups, and customise their profiles.

One of the distinguishing features of Mastodon is its focus on user privacy and moderation. Instance administrators have control over the rules and guidelines for their respective instances, giving users more control over their online experience. Additionally, Mastodon allows users to choose who can see their posts by using different visibility settings, such as public, unlisted, and followers-only.

Overall, Mastodon aims to provide an alternative social media experience that emphasises decentralisation, community moderation, and user control over their data and online interactions. As of March 2023, Mastodon has 10M users.

DeSo, 'the decentralised social blockchain' wants us to 'reimagine the world of social'. DeSo is the first layer-1 blockchain built from the ground up specifically to decentralise social media and scale storage-heavy apps to billions of users – and we know that because it's written on their website. Reading their mission gives us hope of a better social media future:

"We're on a mission to create an internet that's creator-led, user-owned, and open to millions of developers around the world to build off one another."

DeSo articulates this dilemma really well:

"Today, social media is even more centralised than the financial industry was, prior to the creation of Bitcoin. A handful of private companies effectively control public discourse and earn monopoly profits off of content that they don't even create.

Meanwhile, the creators who actually produce this content are underpaid, under-engaged, and under-monetised thanks to an outdated ads-driven business model.

In addition to all of this, the ads-driven business model also forces social media companies to keep a walled garden around content created on their platforms, preventing external developers from innovating or building apps on top of it and giving users and creators no choice but to continue using apps that solely they control.

These problems stem from the fact that the data and content created by users today are privately owned by a handful of companies, rather than publicly accessible as an open utility.

Because only a handful of companies have access to the content, only these companies can curate competitive feeds, only these companies can build competitive new features and apps, and only these companies can monetise this content — content that isn't even created by these companies in the first place.

We're stuck in a loop:

1. Users have to use these companies' apps because they have a monopoly on the content.

2. This forces creators into continuing to give their content up to them in order to get reach.

This results in a vicious cycle that continues to empower these companies at the expense of creators and society as a whole.

These companies have managed to create a global network effect around a private pool of content that they solely monopolise." (source: https://docs.deso.org/)

As of mid-2023, DeSo has about 2M users.

As I write this in mid-2023, Meta's response to X (formerly known as Twitter) has just been released. Threads is a new platform in the Meta ecosystem, very similar to X. The response: 30M users within 24 hours. 100M in less than a week, making it the fastest growing app in history (at the time of writing). For context, ChatGPT reached 100 million users in two months, TikTok reached 100 million users in nine months and Instagram reached 100 million users in two and a half years (source: searchenginejournal). The meteoric rise of Threads shows we still have a way to go before we have a more decentralised world. In my opinion, we don't need more centralised big-tech led social platforms. As a marketer though, I have a responsibility to recommend the best platforms for our clients and their brands. And if 100M+ people are living on Threads, then we still need to be part of it – for now.

So, can we do better than the existing world we know? Web3 promises this. It will take time and a lot of un-learning of behaviours, cultures, norms and expectations. And I am confident that we will get there. To quote Sanhita Baruah, "Good things take time, better things take a little longer".

PART 3:
REDEFINING SUCCESS

Chapter 5:
From Hierarchy to Autonomy: Web3's Impact on Structures

'm a bit of a rebel myself so hierarchy, authority and control never really sat well with me. Early on in my career, I experienced so many problems with hierarchy and I'll take you through a few scenarios.

Firstly, the job application process can be brutal sometimes – and it was when I was looking for my first few professional roles. How many times have you seen entry level job advertisements that require five years' experience? It must happen often, because I've seen so many social media memes created mocking this impossible situation. And how many times have you worked with someone who lied on their resume? I've worked with people who lied about having a three-to-five-year university degree! And there's hardly a verification process for most roles, making it completely possible for people to get away with these lies.

Then there's starting in an entry level position. Depending on the role and company it's not often that people in these roles can offer big contributions or ideas that are taken seriously by senior managers.

Culture is another hot topic. I've spoken to large multi-billion-dollar, multi-national organisations listed on the New York Stock Exchange, who are still developing their company culture in a boardroom, as a strategy, with only senior members of management involved – and then launching this into the business.

A company's culture is not something developed in a boardroom. It is my belief that company culture is the way a team does things, operates, their behaviours, belief system and values – and it takes everyone to exude this behaviour. A team will emulate the culture and behaviours that are shown to them from top management, and this has less to do with what's fabricated in a boardroom by a non-representative section of the company than just actually living and breathing it authentically.

The hierarchical system that most traditional corporate offices follow stifles creativity and ideas. As is often the case, senior management make the majority of decisions and those further down the food chain are usually not heard, or do not feel like they can speak up, contribute and make a change. Of course, this depends on the company and culture itself, but in my experience of corporate life this is the norm.

Blockchain and Web3 alleviates these challenges. Firstly, blockchain allows for verification and secure storge of personal data – so authenticating that people are really who they say they are (and have the qualifications they claim to have) is easier.

In the Web3 world, we have DAOs – Decentralised Autonomous Organisations. Investopedia has an official definition:

"A DAO is a decentralised autonomous organisation, a type of bottom-up entity structure with no central authority. Members of a DAO own tokens of the DAO, and members can vote on initiatives for the entity. Smart contracts are implemented for the DAO, and the code governing the DAO's operations is publicly disclosed."

To put this simply, it's an organisation that operates through smart contracts and blockchain without the need for traditional centralised control or intermediaries. A DAO is designed to be autonomous and self-governing, with decision-making power and governance rules encoded in its code and executed automatically.

In a DAO, members can contribute resources such as cryptocurrency, tokens, or other assets. These contributions give them voting rights and influence over the decisions and actions of the DAO. Voting and decision-making processes within a DAO are typically conducted through a consensus mechanism, where participants can vote on proposals or changes to the organisation's rules and policies.

One of the key features of a DAO is its transparency, as all transactions and decisions are recorded on the blockchain, making them publicly visible and auditable. This transparency helps to ensure accountability and reduces the need for trust in centralised authorities.

DAOs can be used for a wide range of purposes, including decentralised governance, investment funds, crowdfunding, and the management of digital assets. They provide a way for individuals or groups to collaborate, make decisions, and govern themselves in a decentralised and trustless manner.

Overall, they do solve a lot of the issues with traditional corporate centralised structures. Are they perfect? No, but they are a step in the right direction.

Chapter 6:
Ownership Redefined: Tokenomics and Web3 Economy

Tokenomics, my friend, is a term that gets thrown around a lot in the Web3 world. It's all about the economics of tokens in cryptocurrencies and blockchain-based projects. Now, why is this important? Well, understanding tokenomics is crucial if you want to gauge the growth potential of a cryptocurrency or project. It's like peering into the engine of a car to see if it's got the power to go the distance.

Tokenomics is a fundamental component of crypto and blockchain-based economies. It encompasses a wide range of factors that shape how tokens behave and their overall value. Effective tokenomics align the interests of various participants and aim to create a thriving economy by ensuring token utility, scarcity, liquidity, and stability. Just keep in mind that each cryptocurrency and project has its own unique set of rules and principles.

So, let's break it down. Tokenomics encompasses a bunch of factors that govern the behaviour and value of tokens in a digital economy. First up, we've got token distribution. How tokens are initially divvied up and spread around is a big deal. This can happen through things like initial coin offerings (ICOs), token sales, airdrops, or even mining. The way tokens are distributed can have a major impact on the ecosystem. It affects things like decentralisation, who owns the majority of tokens, and how involved the community gets to be.

Another piece of the tokenomics puzzle is token supply. This refers to the total number of tokens in existence. Now, token supply can be fixed, capped, or even inflationary. Here's the thing: our traditional fiat currency system has an unlimited supply, and that often leads to all sorts of challenges. But with tokens, it's a different story. A fixed supply can create scarcity, potentially driving up the value of tokens over time. On the flip side, an inflationary supply means more tokens are being circulated, and that can affect their value in a different way.

Token utility is another key aspect. Tokens serve a specific purpose within the ecosystem. They can be used for things like transactions, governance, staking, accessing services, or even as rewards. The more useful and widely adopted a token is, the higher its value can climb. It's like having a versatile tool in your pocket that opens doors and unlocks opportunities within the digital realm.

Now, let's talk token burning. No, we're not talking about setting tokens on fire (that wouldn't be very productive). Token burning is when tokens are permanently removed from circulation. This can happen through transaction fees, network activity, or even regular token buybacks. The purpose of burning tokens is to decrease the overall supply and increase scarcity. It's a way to manage inflation, enhance token value, and maintain long-term stability.

Governance is another important aspect of blockchain projects. Token holders often have the power to participate in decision-making through voting mechanisms. The more tokens you hold, the more weight your vote carries. This democratic approach to governance encourages community involvement and gives participants a sense of ownership within the ecosystem.

Incentives and rewards also play a crucial role in tokenomics. Projects often use mechanisms to incentivise certain behaviours from participants. For example, you might be rewarded with tokens for mining, staking, or contributing to the network in other ways. These incentives drive engagement and help the ecosystem grow by encouraging users to actively participate and contribute their resources.

Of course, market forces also have a say in tokenomics. The value and trading volume of tokens are influenced by factors like utility, project development, partnerships, regulatory changes, and even market sentiment. Price fluctuations are often driven by how people feel about the overall market and what they think the future holds. Understanding market dynamics and the impact of external factors is key to managing token value and ensuring a stable and sustainable ecosystem.

Now, it's important to note that different projects may have different economic models. Some may opt for deflationary models, where the token supply gradually decreases over time, potentially leading to value appreciation. Others may use algorithmic stablecoins to maintain price stability through supply adjustments, ensuring that the value of the token remains relatively constant.

Let's delve into some real-world examples to put all of this into context. One cryptocurrency that often comes up in discussions about tokenomics is Bitcoin, otherwise known as digital gold. Bitcoin's tokenomics have been widely praised for their design and economic principles. So, let's take a closer look at how Bitcoin's tokenomics work.

First and foremost, Bitcoin has a limited supply. There will only ever be 21 million Bitcoins in existence. This scarcity is a fundamental aspect of its design, ensuring that it cannot be inflated or devalued by arbitrary increases in supply. The limited supply is achieved through a process called mining.

Bitcoin mining involves using powerful computers to solve complex mathematical problems, which validates and secures transactions on the network. Miners compete to solve these problems, and the first one to find a solution is rewarded with newly minted Bitcoins as well as transaction fees. This process adds new Bitcoins into circulation and is how the initial supply of Bitcoins was gradually introduced.

But here's where it gets interesting. Approximately every four years, or after every 210,000 blocks, a so-called halving event occurs in the Bitcoin network. During a halving event, the reward given to miners for successfully mining a block is reduced by half. This mechanism helps control the rate of new Bitcoin issuance and leads to a decreasing inflation rate over time. At the time of writing, the most recent halving occurred in May 2020, reducing the block reward from 12.5 to 6.25 Bitcoins.

Bitcoin's tokenomics also rely on decentralised governance. There is no central authority or entity controlling Bitcoin. Decisions regarding upgrades, improvements, or changes to the protocol are made through a consensus mechanism, where participants signal their support or opposition to proposed changes. This decentralisation is a key aspect of Bitcoin's tokenomics and contributes to its robustness and resilience.

Bitcoin is often considered a store of value, similar to gold. Its limited supply, decentralisation, and strong security make it attractive to individuals seeking a hedge against inflation and a decentralised alternative to traditional financial systems. The value of Bitcoin is determined by market demand and supply dynamics. As more people adopt Bitcoin and recognise its potential, the demand increases, driving up the price.

However, it's important to note that Bitcoin's price is known for its volatility. It can fluctuate significantly in short periods. This volatility is influenced by various factors, including market speculation, regulatory developments, macroeconomic events, and investor sentiment. While Bitcoin has gained substantial value over time, it can also experience significant price corrections.

Bitcoin is just one example of tokenomics in action. There are countless other cryptocurrencies and blockchain projects with their own unique tokenomics. Each project has its own distribution model, token supply, utility, and governance structure. It's a vast and diverse ecosystem with endless possibilities.

Now let's take a look at Dogecoin. Dogecoin, created in December 2013 as a light-hearted cryptocurrency, has a unique tokenomics structure that sets it apart from other cryptocurrencies. Here's a description of the tokenomics of Dogecoin:

1. **Supply:** Dogecoin's token supply is uncapped, meaning there is no maximum limit to the number of Dogecoins that can be created. As of mid-2023, there were over 140 billion Dogecoins in circulation. However, it's important to note

that the Dogecoin blockchain rewards a fixed amount of new Dogecoins to miners per block, resulting in a relatively steady and predictable inflation rate.

2. **Inflation:** Dogecoin has an inflationary token supply model. Unlike Bitcoin, which has a deflationary model with a capped supply, Dogecoin's blockchain rewards a fixed number of new Dogecoins per block indefinitely. Initially, the block reward was set at 10,000 Dogecoins, but it reduces over time.

3. **Block Time:** Dogecoin has a relatively fast block time of approximately one minute. This means that new blocks are added to the Dogecoin blockchain approximately every minute, allowing for faster transaction confirmations compared to some other cryptocurrencies.

4. **Transaction Fees:** Dogecoin transactions typically have low fees due to the relatively low cost of processing transactions on its blockchain. However, during times of high network congestion or increased demand, transaction fees may increase.

5. **Community and Distribution:** Dogecoin has gained significant popularity due to its vibrant and enthusiastic community. It has been used for various charitable causes and tipping content creators. Dogecoins are distributed through mining, where miners contribute computing power to secure the network and process transactions. Additionally, Dogecoin can be purchased on various cryptocurrency exchanges.

But let's go even broader than that. Because Web3 has the potential to revolutionise economic models and reshape the way we interact, transact, and participate in online ecosystems. Here are some key aspects of the potential economic models in Web3:

1. Decentralisation and Disintermediation: Web3 allows for peer-to-peer interactions without the need for intermediaries such as centralised platforms or financial institutions. This disintermediation can eliminate unnecessary middlemen, reduce transaction costs, and increase efficiency in economic activities.

2. Tokenisation and Cryptoeconomics: Web3 introduces the concept of tokens, which can represent various digital and real-world assets, ownership rights, or utility within a decentralised network as described earlier. Tokenisation enables new economic models, such as decentralised finance (DeFi), where users can lend, borrow, and trade digital assets directly without intermediaries. Cryptoeconomics studies the incentive mechanisms that govern these decentralised networks, aligning the interests of participants and ensuring the stability and security of the system.

3. Smart Contracts and Programmable Money: Web3 platforms leverage smart contracts, self-executing contracts with predefined rules and conditions encoded on the blockchain. Smart contracts enable automation, trust, and transparency in economic interactions. They facilitate new models like DAOs where decision-making and governance are collectively managed by token holders, removing the need for traditional hierarchical structures. Programmable money and CBDCs are controversial topics that are still evolving.

4. Open and Permissionless Innovation: Web3 encourages open-source collaboration and permissionless innovation. Anyone can participate, contribute, and build applications on decentralised networks, fostering a more inclusive and diverse economy. This openness allows for the creation of decentralised applications (dApps) that can provide services ranging from decentralised marketplaces to content sharing platforms, with new monetisation and incentive models.

5. Web3 Monetisation Models: Web3 introduces alternative monetisation models beyond traditional advertising and data monetisation. With tokens and smart contracts, content creators can directly monetise their work through micropayments, subscriptions, or other mechanisms. Users can earn tokens by contributing value to the network, such as providing computing power, validating transactions, or creating and curating content.

6. Ownership and Digital Rights: Web3 offers opportunities for individuals to assert ownership and control over their digital assets and data. Through decentralised identity systems, users can manage their digital identities, control access to personal information, and selectively share data with consent. Digital rights management can be more transparent, ensuring fair compensation for creators and protecting intellectual property.

Tokenomics is a vital aspect of crypto and blockchain-based economies. It encompasses various factors such as token distribution, supply and demand dynamics, utility, governance, and incentives. Effective tokenomics align the interests of participants and aim to create a sustainable and thriving economy within a digital ecosystem. By understanding tokenomics, you'll be better equipped to navigate the world of cryptocurrencies and make informed decisions about their potential for growth and value.

Overall, Web3 has the potential to reshape economic models by promoting decentralisation, disintermediation, tokenisation, smart contracts, open innovation, and new monetisation models. It empowers individuals, promotes economic inclusivity, and enables trust and transparency in online transactions, opening up exciting possibilities for a more equitable and decentralised digital economy.

None of the information in this book is to be used as financial advice. I'm simply providing some examples of tokenomics for education purposes only. I hope the information provided puts into perspective the importance of tokenomics.

I want to share a personal story that I experienced in relation to this topic too. While building the primary team for Take3, I undertook many interviews and introductions (more on this later). There are many people in this space, and not all for the right reasons. We have seen some of those 'bad actors' who have made huge errors that impacted millions of people, including Sam Bankman-Fried from FTX and Don Kwon from Terraform Labs. Thankfully though, the majority of people I've connected with in this space are here with good intentions to help shape a more decentralised future.

When I interview people for Web3 roles, one question I often ask is "why Web3? And what does it mean to you?". I get a range of answers, and what I'm looking for is genuine passion and purpose. One particular woman that I met in Sydney did not disappoint. Let's call her Mel for privacy purposes. Mel had not been in Australia for very long. In fact, Mel was originally from the Ukraine and was displaced in early 2022 due to the invasion of Ukraine. So, when I asked her "why Web3, what does it mean to you?", she got emotional telling me how she had to flee her country and if not for cryptocurrency and digital assets, she would have had nothing. She was so thankful for crypto and Web3 to still have access to things she needed to start a new life. An incredible story showing the power of decentralised economies, crypto and Web3.

This is the impact that is possible – and needed (perhaps in some parts of the world more than others).

PART 4:
BEYOND THE HYPE

Chapter 7:
The Art of Possibilities: NFTs and Creative Expression

D o you remember where you were when you first heard about NFTs? Or who told you about them?

For me, it was a webinar. I was introduced to NFTs in late 2020 and to be honest, I was a little confused at first. CryptoPunks was the first NFT project I was exposed to. For those who don't know, CryptoPunks was one of the earliest NFT projects ever launched by a company named Larva Labs.

CryptoPunks are 24x24 pixel art images of punky-looking faces. As written on their website:

"10,000 unique collectible characters with proof of ownership stored on the Ethereum blockchain. The project that inspired the modern CryptoArt movement ... the CryptoPunks are one of the earliest examples of a Non-Fungible Token on Ethereum, and were inspiration for the ERC-721 standard that powers most digital art and collectibles." (source: www.larvalabs.com/cryptopunks)

With all these accolades, and each punk priced at thousands of dollars (each!), I couldn't help but think "Why would anyone be paying so much for one of these?".

The website goes on to explain:

"The CryptoPunks are 10,000 uniquely generated charac-ters. No two are exactly alike, and each one of them can be officially owned by a single person on the Ethereum blockchain. Originally, they could be claimed for free by anybody with an Ethereum wallet, but all 10,000 were quickly claimed. Now they must be purchased from someone via the marketplace that's also embedded in the blockchain. Via this market you can buy, bid on, and offer punks for sale."

I understood the scarcity part, although I was still struggling to comprehend why people were paying hundreds of thou-sands of dollars (sometimes even millions) for one these 'punks'. After further research and education, I understood that this was not simply a purchase – it was a movement, a sub-culture. And CryptoPunks (or should I say, Larva Labs) were at the forefront of it.

In 2021 came the NFT craze. The biggest and most prominent NFT collection was the Bored Ape Yacht Club (BAYC). Yuga Labs was behind this NFT success story. After launching in April 2021, BAYC had a significant impact on the internet and NFT scene. Taken from their website:

"BAYC is a collection of 10,000 Bored Ape NFTs—unique digital collectibles living on the Ethereum blockchain. Your Bored Ape doubles as your Yacht Club membership card, and grants access to members-only benefits, the first of which is access to THE BATHROOM, a collaborative graffiti board. Future areas and perks can be unlocked by the community through roadmap activation." (source: https://boredapeyachtclub.com).

Each bored ape is unique and programmatically generated from over 170 artistic traits created by the artist, Seneca. The BAYC hype and popularity was real. Celebrities and prominent figures own an ape, including Justin Bieber, Paris Hilton, Post Malone, DJ Khaled, Jimmy Fallon, Kevin Hart, Snoop Dogg and Madonna. The difference between CryptoPunks and BAYC for me? I understood BAYC better and more quickly. Why? Because BAYC had utility and a roadmap. Being an owner in the BAYC community came with utility in the form of private online spaces, exclusive merch, airdrops and members-only live events. CryptoPunks were more of a collectible or a PFP (profile picture) only.

As I educated myself more, I came to understand the value of NFTs – and its related tech – a lot better. I learnt about the importance of strong utility, or real-world use case. I came to understand that utility is important for an NFT collection because it provides added value and benefits to the NFT collectors and holders. Utility refers to the functionality or purpose of an NFT beyond its basic ownership and provenance. It gives collectors more reasons to acquire and hold NFTs, enhancing their overall experience and potential return on investment. I've summarised the key reasons why utility is important for NFTs:

1. Enhanced Interactivity: NFTs with utility can offer interactive features that engage collectors in unique and immersive experiences. For example, an NFT might grant access to exclusive events, virtual worlds, or games. This interactivity increases the engagement and enjoyment of owning the NFT.

2. Access to Exclusive Content: Utility can provide collectors with access to exclusive content related to the NFT collection. It could include unreleased artwork, behind-the-scenes

materials, or special merchandise. By offering access to exclusive content, the value of the NFT is enhanced.

3. Membership Benefits: Some NFT collections offer membership programs where owning certain NFTs grants holders additional perks or privileges. These benefits could include priority access to future releases, discounts on merchandise, or even governance rights within a community. Membership benefits create a sense of exclusivity and incentivise collectors to acquire and hold NFTs.

4. Royalties and Revenue Sharing: Utility can enable NFT holders to earn ongoing royalties or participate in revenue-sharing mechanisms. For instance, if an NFT represents ownership of a digital artwork or a music track, the holder may receive a percentage of the proceeds every time the artwork is resold or the track is streamed. This financial incentive makes the NFT more attractive to potential buyers. Even more importantly, content creators and artists have suffered for too long from plagiarism and intermediaries taking a cut for their hard work. NFTs/blockchain offer an opportunity for creators to be correctly credited and compensated for their skills.

5. Future Utility Potential: NFTs with utility may have the potential for future enhancements or developments. This can increase the long-term value and demand for the NFTs. For example, an NFT collection might have plans to introduce new features, collaborations, or integrations that would further enhance the utility of the existing NFTs.

Overall, utility adds practical functionality and benefits to an NFT collection, making it more desirable for collectors and potentially increasing its long-term value. However, it's important to note that utility should be carefully designed and executed to ensure it aligns with the goals and vision of the NFT collection and its community.

My personal favourites are membership/loyalty benefits as well as rightful ownership and credit to content creators. That's right, NFTs can also be used for membership and loyalty benefits. How cool is that?

Imagine you're a die-hard fan of a particular brand, artist, or creator. You're always on the lookout for exclusive perks and rewards that come with being a loyal member. NFTs can take that loyalty to a whole new level. They can serve as virtual membership cards or badges that unlock special privileges and experiences.

As a member, you could receive a unique NFT that acts as your virtual ticket to exclusive events, meet-and-greets, or early access to limited-edition merchandise. These NFTs can be securely stored in your digital wallet, and whenever you want to access your membership benefits, you simply present your NFT.

But it doesn't stop there. NFTs can also be used to grant you ownership of virtual assets or experiences. Let's say you're a member of a gaming community. With an NFT, you could own rare in-game items, unique character skins, or even virtual real estate. These digital assets can hold significant value and become highly sought after by other members or collectors.

The beauty of NFTs is that they are easily transferable and verifiable on the blockchain. So, if you decide to sell or trade your membership NFT, you can do so with confidence, knowing that the ownership is transparent and recorded on the blockchain for all to see. This opens up a whole new world of possibilities for buying, selling, and trading membership perks and benefits.

But it's not just about the perks and benefits. NFTs also create a sense of exclusivity and community. Owning a unique NFT that represents your membership can make you feel like part of an elite group, connected to others who share the same passion and dedication. With NFTs, membership becomes more than just a card or a number – it becomes a digital token of belonging, authenticity, and value.

Now, I know what you might be thinking. What about the environmental impact of NFTs? It's true that the energy consumption associated with some blockchain networks can be a concern. However, there are efforts underway to develop more eco-friendly solutions, such as using proof-of-stake algorithms instead of the energy-intensive proof-of-work. It's important to weigh the pros and cons and choose platforms that prioritise sustainability.

Another one of their most powerful applications is in ensuring that creators receive proper recognition and compensation for their work. In the digital age, it's all too easy for content to be copied, shared, and plagiarised without the original creators getting their fair share. But NFTs change the game. These unique tokens are built on blockchain technology, which means they can serve as digital certificates of authenticity and ownership.

When a content creator mints an NFT, they can attach it to their artwork, music, videos, or any other form of digital content they produce. This NFT becomes a digital representation of the original work, verifying its authenticity and attributing ownership to the creator. It's like having a digital signature or a watermark that can't be removed.

By using NFTs, content creators can establish a direct connection with their audience and ensure that their work is properly credited. Whenever the NFT is sold or transferred, a portion of the proceeds can be automatically directed back to the creator through smart contracts. This creates a system where artists can receive royalties for their work even after it's been sold multiple times in the secondary market.

But it doesn't stop at monetary compensation. NFTs also enable creators to maintain control over their work and set specific terms for its usage. For example, they can include licensing agreements within the smart contracts attached to their NFTs, specifying how the content can be used, reproduced, or displayed. This gives creators the ability to protect their intellectual property rights and ensure that their work is used in accordance with their wishes.

Another powerful aspect of NFTs is that they provide a verifiable and immutable record of the entire ownership history of a piece of digital content. This means that even if the work is shared or displayed on various platforms, the NFT serves as a proof of the original creator's contribution and authorship. It's a digital trail that can't be erased or manipulated, providing content creators with the recognition they deserve.

NFTs also foster a closer relationship between creators and their fans. Collectors who purchase NFTs become part of a unique community, supporting and engaging directly with the artists they admire. This creates a sense of connection and appreciation that goes beyond traditional forms of consumption.

NFTs offer content creators a ground-breaking way to establish rightful ownership, receive fair compensation, and get the credit they deserve. These digital tokens provide a secure and transparent mechanism for artists to protect their work, track its provenance, and engage directly with their audience. With NFTs, the digital world becomes a more equitable and rewarding place for creators, ensuring that their artistic contributions are properly acknowledged and valued.

Chapter 8:
Transforming Industries: Web3 Applications Beyond Hype

Web3 is disrupting various industries, and we're just scratching the surface of its potential. It's fascinating to see how this technology is impacting different sectors, from finance to cloud storage, arts, events, publishing, and beyond. While some brands are diving into Web3 with NFT drops, there are also innovators working on revolutionary real-world applications that are truly shaping the future of blockchain.

Let's look at a few notable examples to get a glimpse of what's happening in the space. Keep in mind that I'm sharing these examples for educational purposes only, and none of this should be construed as financial advice.

In the world of arts and collectibles, we have platforms like SuperRare and Rarible. Artists can mint their unique creations as NFTs and sell them directly to collectors, cutting out the middleman and ensuring they receive proper credit and compensation. It's an exciting way for artists to showcase their talent and build a direct connection with their audience.

When it comes to cloud storage, Filecoin is making waves. It's a decentralised storage network that allows individuals and organisations to rent out their unused storage space in exchange for Filecoin tokens. This not only creates a more efficient and secure storage system but also provides an opportunity for users to monetise their unused resources.

In the finance sector, decentralised finance (DeFi) is revolutionising traditional banking and lending. Platforms like Compound and Aave enable users to lend and borrow funds directly from others, removing the need for intermediaries like banks. This opens up new opportunities for financial inclusion and empowers individuals to have greater control over their financial decisions.

Additionally, Web3 is transforming the publishing industry. Platforms like Mirror and Audius are redefining the way content is created, distributed, and consumed. Writers can publish their work directly to the blockchain, ensuring transparency, ownership, and fair compensation. Musicians can share their music without intermediaries, enabling them to retain more control and receive direct support from their fans.

These examples represent just a fraction of the exciting Web3 developments and showcase the potential for decentralisation, transparency, and empowerment across various industries.

Here I dive deeper into some Web3 versions improving upon their Web2 counterparts:

Digital signatures
Web2: DocuSign
Web3: KwikTrust

You probably know or have used DocuSign before – the cloud-based, digital signature solution that allows users to sign and manage documents in a streamlined, efficient way. Particularly during the pandemic, digital signatures become commonplace, which is great for the environment as we limit unnecessary printing. KwikTrust, its Web3 version, uses blockchain tech to deliver an immutable record of transactions, increasing security in the document management process. A unique digital signature is created for each transaction and stored on the chain. As KwikTrust was only founded in 2020, it's working to catch up to DocuSign's integration capabilities with the likes of Salesforce and Dropbox, but certainly offers enhanced security and trust.

Domain name registration
Web2: GoDaddy, Crazy Domains
Web3: Ethereum Name Service (ENS)

While GoDaddy exists as a centralised register for domain names (and additional services such as website hosting), ENS allows users to register readable domain names with an '.eth' extension, which can be used to access dApps and services in the ETH ecosystem. You can find me at reneefrancis.eth (also my Twitter handle). You can also associate a domain name with a wallet address. A decentralised domain register also means more security and control, and easier management of your domain name and associated wallet addresses.

Publishing
Web2: Medium
Web3: Mirror

Open to ETH wallet holders, Mirror launched in December 2020 as a decentralised blogging platform – think Medium, but Web3. Mirror allows users to publish content such as articles, podcasts and videos, and retain the ownership rights to whatever they publish while earning royalties in cryptocurrency. Creators can also monetise their content by selling work as NFTs, allowing other users to 'collect' pieces of interest. The decentralised nature of Mirror means it is more resistant to censorship and curation, allowing royalties to transfer straight to creators.

Music streaming
Web2: Spotify
Web3: Audius

Audius is the decentralised answer to Spotify, still in the early development phases but showing great promise as a more secure and equitable music platform. While Spotify is a centralised company with a traditional payment system, with royalties earnt based on number of streams, Audius elevates opportunities for up-and-coming and independent artists in particular to earn AUDIO tokens for publishing their music to the platform. Any musician can set up a profile and build their fan base. While the interface isn't as user-friendly as Spotify's just yet, users can still tune in to trending songs, as well as discover new music, search for songs based on mood and listen to curated playlists of users they follow. Listeners can further support musicians they like on the platform by purchasing and staking AUDIO tokens, which also grants them access to exclusive content.

Event ticketing
Web2: Ticketek
Web3: UTIX

Another learning through the Investment Mastery webinar: UTIX. Ticketek and similar ticketing agents are based on a centralised platform, with stakeholders like promoters and agents involved in the ticketing process. As such, ticket prices need to accommodate for those stakeholder cuts as well as for the event itself. Ticket sales from these sorts of platforms have also come under scrutiny over concerns around price gouging and ticket hoarding, which UTIX aims to put an end to. Based on a blockchain network, UTIX transactions come with a higher level of security, reducing the risk of scams and ticket scalping. It also means greater transparency in ticket sales as a prede-termined smart contract is set to include ticket price. Without intermediaries seeking a cut from transactions, the idea is that tickets also have a lower price point and a more secure payment process.

Video streaming

Web2: Netflix

Web3: Theta network

While Theta serves as a more cost-effective alternative for video streaming for businesses, users benefit too. At present, most video streaming services rely on a content delivery network (CDN) of multiple servers around the world that help to deliver content to millions of users quickly. CDNs are traditionally expensive, as they charge for data used. This means that if a streaming site is popular, they pay more – and pass that charge on to subscribers. At a very top level, Theta network aims to solve this problem with a blockchain-powered, peer-to-peer video delivery service. This means users can participate in the video delivery process and be rewarded with tokens for assisting. The idea behind Theta isn't really to serve as a competitor to the likes of Netflix or even YouTube, but rather serve as a protocol to help them harness more efficient and cost-effective video distribution and (hopefully) offer their users a more cost-effective subscription service.

One thing you may notice about the above Web3 platforms is the existence of blockchain tech. Security and transparency are key to operations across so many industries, and blockchain offers a 'levelled-up' version of that. Businesses benefit from more streamlined, cost-effective and secure transactions; users benefit from increased purchase security and true data ownership. I could go on with more examples, as there are many, however these are relatable to the apps or websites you might use on a daily basis.

PART 5:
THE FUTURE UNLEASHED

Chapter 9:
Web3 Governance: Power to the Community

Picture this: instead of a few bigwigs making all the decisions, DAOs (decentralised autonomous organisations) distribute power and authority among network participants. It's like a digital democracy where everyone gets a say, no matter where they're from. How cool is that?

DAOs are powered by smart contracts, which are like self-executing agreements written on the blockchain. These contracts govern voting mechanisms, fund allocation, and proposal execution. Token holders have the power to vote on proposals, whether it's deciding on investments or upgrading protocols. It's a democratic process where voting power is often tied to the number of tokens you hold.

One of the greatest strengths of DAOs is transparency. Since all decisions and transactions are recorded on the blockchain, you can bid farewell to shady dealings happening behind closed doors. It's like having a front-row seat to witness every move and hold everyone accountable. Transparency is the name of the game.

DAOs also bring inclusivity to the table. They open doors for anyone to participate. No more being excluded because of where you live or how much money you have. It's all about giving everyone a fair chance to make their voice heard and impact the projects or platforms they care about.

Of course, DAOs face their fair share of challenges. Governance scalability is a big one. As DAOs grow, ensuring that decision-making processes can handle the increased load becomes crucial. Meaningful participation is another challenge. DAOs need to find ways to engage and involve their token holders effectively. Nobody wants a community where people just sit on the sidelines.

Then there's the issue of sybil attacks. These sneaky attacks happen when individuals create multiple identities to gain more voting power. DAOs need to be on guard and find ways to tackle this problem to maintain the integrity of the decision-making process. And let's not forget about handling contentious decision-making. When opinions clash, finding consensus can be a real challenge.

But don't let these hurdles dampen your enthusiasm because the potential of DAOs is mind-boggling. They can revolutionise finance, governance, art, and even charitable organisations.

In the world of finance, DAOs are turning traditional systems upside down. They enable decentralised lending, crowdfunding, and investment platforms. No more relying on banks or traditional intermediaries. DAOs make it possible for capital to flow more efficiently, making financial systems more inclusive and empowering individuals.

When it comes to governance, DAOs are like the heroes of democracy. They bring transparency and accountability to the decision-making process, making it more participatory and responsive. Imagine having a real say in the policies that affect your life, whether it's within organisations or even at governmental level. DAOs are leading the way to a more democratic future.

DAOs are unleashing a creative revolution in industries like art and music. Artists can tokenise their work, sell directly to fans, and receive fair compensation without intermediaries taking a big chunk of the pie. It's a direct connection between artists and their audience, bypassing the gatekeepers. It's a game-changer for creators, empowering them to thrive on their own terms.

And let's not forget about social impact. DAOs are not just about making money; they're about making a difference. They enable communities to come together, fund charitable projects, support environmental initiatives, and tackle social injustices. It's like a global network of superheroes joining forces to create a better world.

DAOs are rewriting the rules of the game by embodying the principles of decentralisation, transparency, and inclusivity, giving power back to the people. While they face challenges, their potential to revolutionise industries and empower communities is immense. DAOs are like the champions of collaboration and coordination. It's like having a worldwide team that works together seamlessly. No more shady behind-the-scenes stuff! DAOs are all about fairness and giving everyone a voice.

According to Legal Nodes, in 2023 the best countries for implementing a DAO include Marshall Islands, the US (Wyoming), Switzerland, the Cayman Islands, Liechtenstein, Singapore, Panama, the British Virgin Islands, Gibraltar, and the Bahamas.

Let's explore an example of a DAO. Uniswap DAO, a decentralised autonomous organisation that's shaking up the crypto scene. Uniswap DAO is like the boss of the Uniswap protocol, which is one of the hottest decentralised exchanges (DEX) out there. It lets you swap cryptocurrencies without dealing with middlemen. But here's the kicker: Uniswap DAO isn't just about trading—it's also all about community power and decentralised decision-making. In this overview, we'll break down the key bits, the governance stuff, and the rad accomplishments of the Uniswap DAO.

The Uniswap protocol is a game-changer in the world of decentralised trading. Forget those old-school exchanges with order books. Uniswap does things differently. It uses liquidity pools and smart contracts to let you trade tokens directly with others. So, instead of relying on a centralised exchange, you're interacting with a network of users. Liquidity providers are the heroes here—they throw their funds into these pools and earn fees based on how much liquidity they provide. Uniswap uses some fancy maths called the constant product market maker (CPMM) algorithm to figure out the token prices within the pools.

Let's talk about how Uniswap DAO came into existence. It all started with the launch of Uniswap V3 in May 2021, and things got a lot more exciting. Uniswap DAO was born, adding a whole new level of community-driven power to the protocol. The goal? To let token holders and the wider community have a say in shaping the future of Uniswap and to help it grow and thrive in the long run. That's where the DAO structure comes in. Uniswap DAO lets the community govern the protocol, making decisions together and keeping things decentralised.

So, how does Uniswap DAO actually work? Well, it's all about giving the power to the people. If you're holding Uniswap's native governance token, UNI, you're in the game. UNI token holders can propose ideas, discuss them with the community, and vote on important matters. The more UNI tokens you have, the more weight your vote carries. Of course, they've set some rules to keep things fair. You need to have a minimum amount of UNI tokens to create proposals and cast votes. This way, they prevent spam and ensure that serious players are involved.

To make the governance process smooth and easy, Uniswap created the Uniswap Governance Interface. It's a user-friendly platform that lets you join the discussion, show your support for proposals, and vote using on-chain transactions. They really wanted to make sure that everyone has a chance to be heard and make an impact.

Uniswap DAO has achieved some awesome things. First off, in 2020, they did something pretty cool and launched the UNI token through an airdrop. And guess what? They gave it to past users of the Uniswap platform. Talk about spreading the love! This move got the community excited and engaged, and it was a big step towards making Uniswap more decentralised.

Uniswap DAO has also been a driving force behind many protocol upgrades and improvements. These upgrades cover stuff like fees, governance parameters, and scalability. And the best part? These changes were driven by the community. They proposed ideas, discussed them, and voted to make them happen.

Uniswap DAO has also been a magnet for partnerships and joined forces with other projects to create a DeFi dream team. These partnerships have made Uniswap even more awesome, allowing users to do all sorts of fancy things by combining different protocols. This has solidified Uniswap's position as a leading DEX in the world of decentralised finance.

People love using Uniswap because it offers high liquidity and attracts a diverse range of projects and users. It's become the go-to spot for trading and swapping tokens in the DeFi universe.

Uniswap DAO is like the cool kid in the blockchain neighbourhood. It's revolutionising decentralised trading while giving the power back to the community. No more dealing with middlemen or relying on centralised exchanges. With Uniswap DAO, you're part of something bigger. You can swap tokens directly, propose ideas, and vote on important decisions. It's all about inclusivity and making sure everyone has a say.

Uniswap is just one example of a DAO implementation that is doing a good job. Not an easy feat while being so early in the game.

A bunch of people coming together, making decisions, and running things without any central authority? That's what DAOs are all about. They're like a community on steroids, using blockchain technology to collaborate and govern themselves. With DAOs, power is spread out, and decisions are made by the people, not just some big shots. It's a revolution in how organisations operate, opening up new possibilities and empowering individuals.

Chapter 10:
Building Web3: Challenges and Opportunities

We've explored how Web3 is all about decentralisation, blockchain technology, and a user-centric internet experience. But before we get too excited, let's take a closer look at the hurdles we face, the scalability issues we encounter, and why user adoption is crucial in this space.

Firstly, let's talk technical hurdles. Building in Web3 is no walk in the park. We're talking about a paradigm shift that disrupts the traditional way the web and the world operates. The primary challenge lies in decentralisation. Web3 aims to eliminate the need for centralised authorities by relying on distributed networks and blockchain technology. This decentralised approach introduces a whole new set of technical complexities.

One of the major hurdles is interoperability. Web3 envisions a future where different blockchain networks can seamlessly communicate with each other. However, achieving this interoperability is easier said than done. Different blockchains employ different consensus mechanisms, programming languages, and data structures. Bridging these gaps and creating an interconnected ecosystem requires significant technical expertise and standardisation efforts.

Scalability is another thorny issue. The current state of many popular blockchains, like Ethereum, shows that scalability can be a bottleneck. As more users and applications join the network, the transaction throughput becomes constrained, leading to higher fees and slower confirmation times. Solving this scalability challenge is crucial for the widespread adoption of Web3. Luckily, various scaling solutions, such as layer 2 protocols, sharding, and sidechains, are being actively developed to tackle this.

Now, let's address the elephant in the room: user adoption. No matter how technically sophisticated Web3 becomes, it won't reach its true potential without user adoption. We need to make it easy and seamless for average users to interact with decentralised applications (dApps) and blockchain-based services. This requires creating intuitive user interfaces, improving user experience, and educating users about the benefits and possibilities of Web3.

One major hurdle to user adoption is the steep learning curve. Blockchain technology and the concepts behind Web3 can be intimidating for newcomers. To bridge this gap, we need to focus on providing user-friendly tools and interfaces that abstract away the complexities. Imagine if using Web3 felt as easy as using popular social media platforms or online banking. We need to design dApps with user-centricity in mind, ensuring that the benefits of decentralisation are easily accessible and understandable to everyone.

Addressing scalability issues is also crucial for attracting users. High transaction fees and slow confirmation times are major turn-offs for potential users. Imagine trying to use a decentralised finance (DeFi) application only to be deterred by exorbitant fees that outweigh the benefits. Improving scalability through innovative solutions will not only enhance the user experience but also make Web3 more economically viable for a wider audience.

To encourage user adoption, we must also address privacy and security concerns. Blockchain technology has the potential to revolutionise data ownership and privacy, but it also introduces new risks. Users need to feel confident that their personal information and digital assets are secure. Robust security measures and privacy-enhancing technologies should be at the forefront of Web3 development to foster trust and attract users.

In the next chapter, I explore the six biggest challenges I believe we need to overcome for mainstream user adoption.

In addition to technical and usability challenges, regulatory and legal considerations play a significant role in shaping the future of Web3. As decentralised applications and blockchain networks gain prominence, governments around the world are grappling with how to regulate this emerging field. Striking the right balance between innovation and regulation is crucial for ensuring a thriving Web3 ecosystem while safeguarding against fraudulent activities and protecting users' rights.

Despite the challenges, the opportunities presented by Web3 are vast. Web3 has the potential to reshape industries, empower individuals, and create a more inclusive and equitable digital landscape.

In conclusion, building Web3 is a monumental task that comes with its fair share of challenges. Overcoming technical hurdles, addressing scalability issues, and focusing on user adoption are all critical steps in realising the vision of a decentralised and user-centric internet. By tackling these challenges and embracing the opportunities presented by Web3, we can pave the way for a more decentralised, secure, and empowering digital future. So, let's roll up our sleeves and build the Web3 we envision together. It won't be easy, but it will definitely be worth it.

Chapter 11:
Overcoming Challenges for Mainstream User Adoption

There are three certainties in life: death, taxes, and web 3. This new iteration of the internet has brought with it an ever-growing community of visionaries, investors, entrepreneurs and events. Locally, Australia's Web3 community has made itself known through monthly community meet-ups and big-ticket events such as the Australian Crypto Convention, Blockchain Week and NFT Fest. According to an October 2022 report from Finder, Australia ranks fifth out of 26 countries for crypto adoption. Even with all this growth, web 3 continues to face its challenges on the path to mainstream adoption. Here are some of the key roadblocks in my opinion.

1. **The Innovation adoption curve**
   Ah, the journey of adoption. When it comes to Web3, we're still at the early stages of the game. You know how it goes - there are always those trailblazing early adopters who

jump on first. But the real challenge lies in bridging the gap between those early adopters and the early majority.

According to Geoffrey Moore's theory, this gap is called 'crossing the chasm.' It's that pivotal moment when a new technology transitions from being embraced by the enthusiasts to gaining widespread adoption. And guess what? We're not quite there yet in the world of Web3.

Let's take a trip down memory lane and look at the early days of the internet. It's hard to believe, but it took many years for the internet to become mainstream. Back in 1995, only a tiny 0.4% of the global population was using the internet. By 2000 that number was just 5.8%. It wasn't until July 2022 that we finally reached a whopping 69% global internet adoption (source: ITU) That's a journey of 27 years to get to that level. Now, let's put things in perspective - Bitcoin and Blockchain were born just 15 years ago. We're still relatively early in the game.

To cross this chasm and achieve mainstream adoption in Web3, education initiatives and marketing activations will play a crucial role. We need to educate and influence the early majority, those folks who are eager but still hesitant to fully embrace this new wave of technology. By providing clear and accessible information, we can help them understand the benefits, possibilities, and potential pitfalls of Web3.

Education is key. We need to break down the complex concepts, demystify the jargon, and make Web3 relatable and understandable to the everyday person. It's about showing people how Web3 can enhance their lives, whether it's through better data ownership, new economic opportunities, or decentralised platforms that put users in control.

But it's not just about education. We also need strategic marketing activations that capture the attention and imagination of the early majority. We need creative campaigns that showcase the real-world applications of Web3, highlighting success stories and demonstrating how it can make a tangible difference. By showing people what's possible, we can inspire them to take that leap and embrace this exciting new frontier.

As the early majority starts to embrace Web3, the ripple effect will naturally occur. The late majority and even the laggards will gradually follow suit. But our focus for now should be on winning over that crucial early majority. It's like planting seeds in fertile soil - once they take root and flourish, the rest will follow.

It may take time, but with the right mix of education and marketing efforts, we can cross that chasm and bring Web3 to the masses.

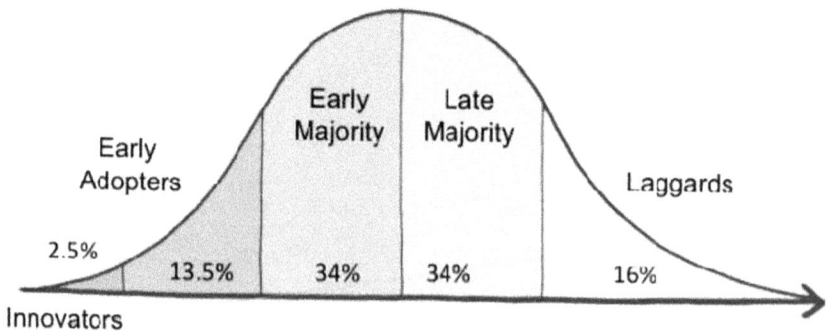

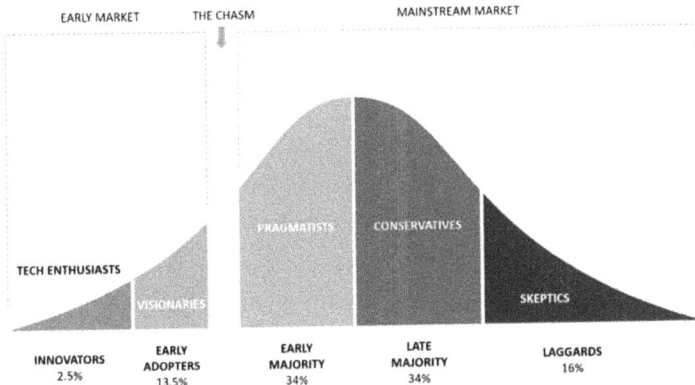

EARLY MARKET   THE CHASM   MAINSTREAM MARKET

TECH ENTHUSIASTS   PRAGMATISTS   CONSERVATIVES

VISIONARIES   SKEPTICS

| INNOVATORS 2.5% | EARLY ADOPTERS 13.5% | EARLY MAJORITY 34% | LATE MAJORITY 34% | LAGGARDS 16% |

2. **Education**

A lot of people are still stuck in the mindset of believing that it's acceptable for global tech giants to own and control our data. We've been conditioned to accept that they're the ones who dictate how our information is used. But here's the thing - with something as innovative and disruptive as blockchain, education becomes key to driving adoption.

In recent years, we've seen controversies surrounding platforms like Meta (formerly known as Facebook). These incidents have made us question whether there's a better way. And you know what? Web3 provides that alternative. Instead of paying companies for our personal data, it makes more sense to cut out the middleman and pay users directly through tokens. It's a shift in mindset that aligns with the principles of decentralisation and empowers individuals to take control of their own data.

However, let's not underestimate the challenge ahead. Moving to Web3 means asking people to un-learn what they've been taught about currency, data ownership, centralised structures, control, community, and even social media. It's like rewiring our brains and reimagining the way we interact with the digital world. It's not an impossible task, but it definitely presents some obstacles.

To make this transition smoother, education becomes paramount. We need to provide clear and accessible information to help people understand the benefits and possibilities that Web3 brings. We need to explain how blockchain technology works, how data ownership can be decentralised, and how communities can have a say in decision-making processes. It's about breaking down complex concepts and making them relatable and easy to grasp.

Moreover, we need to showcase real-world examples and success stories that demonstrate the value and potential of Web3. By highlighting how individuals can monetise their own data, participate in decentralised platforms, and be part of vibrant communities, we can inspire others to embrace this new way of interacting with the digital world.

It's also important to address the concerns and fears that people may have. Change can be intimidating, and Web3 represents a significant shift from the familiar landscape of Web2. We need to acknowledge these concerns and provide reassurance. We can't expect everyone to jump on board immediately, but with patience, understanding, and continuous education, we can help more people see the benefits and possibilities of Web3.

Education is the key to driving adoption of Web3. We need to break free from the conditioned mindset of global tech giants controlling our data and embrace a more decentralised future. By un-learning old habits and opening our minds to new possibilities, we can lead the way to a more equitable and user-centric digital landscape.

3. **Rug pulls and scams**

Let's talk about something that's unfortunately all too common in the online world (and not just Web3): scams. They've been around since the early days of the internet, but when it comes to Web3, they seem to be rampant. These scams prey on people's hesitancy, fear, uncertainty, doubt, and resistance. One particularly nasty scam is called a rug pull, where a project developer or founder promotes a new project, collects investors' money, and then disappears. It's no wonder that these rug pulls can easily scare people away from Web3. After all, once bitten, twice shy, right?

To make matters worse, recent controversies surrounding projects like LUNA and FTX have only added fuel to the sceptics' fire. These incidents have given them more reasons to doubt the legitimacy and credibility of crypto and Web3 as a whole. It's just extra ammunition to shoot down the idea of embracing this new frontier.

But here's the thing: education is the key. With the right guidance and support, people can learn how to avoid rug pulls and scams. They can gain the knowledge and skills to navigate the Web3 space safely and make informed decisions. Education empowers individuals to protect themselves and thrive in Web3.

It's important to remember that not every project or platform in Web3 is a scam. There are plenty of legitimate and innovative projects that are driving real progress and offering genuine value. By educating ourselves and staying informed, we can differentiate between the good and the bad, the trustworthy and the shady.

For those looking to explore Web3, doing your due diligence is crucial. Research the projects you're interested in, read reviews and feedback from trusted sources, and seek out communities and forums where experienced members share their insights. Surrounding yourself with a supportive and knowledgeable network can make all the difference in navigating the sometimes murky waters of Web3.

As the Web3 community, we have a responsibility to promote transparency, accountability, and best practices. By calling out scams and sharing information about them, we can help protect others - it's about looking out for one another and fostering an environment of trust and support.

While scams may be an unfortunate part of the online world, they shouldn't define the Web3 landscape. By arming ourselves with knowledge, staying vigilant, and working together, we can create a Web3 ecosystem that thrives on trust and genuine innovation.

Remember, the power is in our hands to prosper in the Web3 space and we should not let the fear of scams hold us back. With the right mindset and a commitment to learning and growth, we can embrace this exciting new frontier and reap its benefits.

And remember, getting scammed or caught up in a rug pull in the world of Web3 is like a wild initiation ceremony you never signed up for. It's that unexpected, 'Welcome to the club!' moment where you realise the darker side of this exciting new frontier. It's a rollercoaster ride that starts with anticipation and ends with a stomach-churning drop. While it may not be the most pleasant experience, it's a valuable lesson that opens your eyes to the importance of education, research, and community support in this space. Consider it a rite of passage that toughens you up and prepares you to navigate the wild, wild Web3 with a sharper eye and a healthier dose of scepticism.

4. The 'IOU' of NFTs

Let's talk about the influx of NFT projects that are or have launched as 'IOUs' in the Web3 space. Now, I don't know about you, but this trend doesn't exactly inspire confidence in me. I mean, asking people to invest their hard-earned money with the promise of some future value and utility? That's a risky move for both investors and project owners.

Sure, I get it. The idea behind these IOU projects is that they'll eventually offer something valuable and useful down the line. But let's be real here - promises don't always translate into reality. And that's where the problem lies. While some of these IOU projects might be legitimate, they also open the door for scams and rug pulls. It's like walking a tightrope blindfolded, hoping you won't fall into the abyss.

If we want Web3 to thrive and gain credibility, builders and visionaries in this space need to come armed with real solutions and real value upfront. We need projects that tackle the problems of the current Web2 world. The more tangible and practical these projects are, the better chance we have of evolving Web3 with stability and credibility.

Sadly, the allure of making huge profits quickly can attract individuals with less-than-honourable intentions into the Web3 space. These opportunistic folks might not care about building something substantial or contributing to the greater good – they're just looking for a quick buck, and that's not what we need to make Web3 a trustworthy and welcoming environment.

To overcome this challenge, we need to shift our focus towards genuine projects that create real value. Instead of being blinded by the promise of instant riches, let's prioritise projects that address real problems and offer practical solutions. That way, we can create an environment that fosters trust and encourages people to embrace Web3.

Building a solid foundation of trustworthy and impactful projects is crucial for Web3's long-term success. We want to show the world that this space is more than just a get-rich-quick scheme. It's a place where innovation, creativity, and genuine solutions come together to create a better digital future. Together, we can make Web3 a place we can all believe in.

5. User experience

We're about to dive into the world of user experience (UX) and why it's essential to level up the game in Web3. While the concept of Web3 is exciting, we can't ignore the fact that the user experience has a big role to play in attracting the masses. Let's talk about why UX matters and how we can make it more user-friendly and accessible.

First off, let's address the elephant in the room: the learning curve. Web3 and its underlying technologies can be quite complex and intimidating for newcomers. If we want Web3 to reach the masses, we need to make it user-friendly and easy to understand. We have to bridge the gap between the technology and the user's experience.

One way to achieve this is by designing intuitive user interfaces (UI) that hide the technical complexities behind the scenes. Imagine if using Web3 felt as familiar as using your favourite social media platform. By simplifying the process, we can encourage more people to explore the world of Web3 without feeling overwhelmed.

Another crucial aspect of UX in Web3 is seamless onboarding. When users first step into Web3, they should be guided through the process smoothly and with clear instructions. Clear and concise explanations, tooltips, and interactive tutorials can help users navigate the new terrain with confidence. First impressions matter, and a positive onboarding experience can make a world of difference in attracting and retaining users.

We also need to address the issue of transaction fees and confirmation times. Let's face it; high fees and long confirmation times can be major turn-offs for potential users. Imagine trying to buy or sell something on a Web3 marketplace, only to be greeted with sky-high fees that make you reconsider. To gain mainstream adoption, we must prioritise scalability solutions and optimise transaction processing to make it faster and more affordable.

As with any new concept, education plays a critical role in improving the user experience. Web3 is still relatively new, and many people are not aware of its potential or how to utilise it effectively. By providing comprehensive educational resources, tutorials, and guides, we can empower users to explore Web3 confidently. Education should not only cover the technical aspects but also emphasise Web3's benefits and possibilities.

Privacy and security are also paramount in gaining mainstream adoption. Users need to feel confident that their personal information and digital assets are safe and secure. Robust security measures, such as secure wallets and encryption techniques, need to be implemented to build trust and ensure the safety of user data.

Collaboration and community involvement are essential ingredients in improving the Web3 user experience. Developers, designers, and users should work hand in hand to gather feedback, iterate on designs, and co-create a better Web3. By actively listening to user needs and incorporating their insights into the development process, we can build a Web3 that truly meets their expectations and requirements.

Lastly, compatibility and integration with existing Web2 platforms should be a priority. Instead of treating Web3 as a separate and secluded realm, we should focus on creating seamless interactions between Web2 and Web3. Integrating Web3 features into existing platforms and applications can help users gradually transition into the world of decentralisation without feeling like they have to abandon what they already know.

Leveling up the user experience is crucial for Web3 to gain mainstream adoption. By focusing on intuitive interfaces, seamless onboarding, affordability, education, privacy, security, collaboration, and compatibility, we can create a more user-friendly and accessible Web3 ecosystem.

6. **The expectation that Web3 will replace Web2 entirely**

First things first, Web3 is not meant to replace Web2; it's designed to work alongside it. Web2, also known as the traditional internet we use today, has served us well with its centralised platforms and familiar user experiences. However, Web3 takes things to a whole new level by incorporating decentralisation and blockchain technology.

To better understand how Web3 can complement Web2, let's look at a real-world example. Nike, a well-established and solid Web2 brand, made a strategic move in December 2021 by acquiring RTFKT, a leading Web3 collectibles brand. Nike recognised the potential of Web3 and saw an opportunity to leverage RTFKT's expertise in the space to aid in their digital transformation. This acquisition showcased how Web3 and Web2 can work together harmoniously.

Integrating Web3 elements into existing businesses and platforms is key to driving adoption and demonstrating the value of this new paradigm. By embracing Web3, companies can enhance user experiences, create new revenue streams, and tap into the growing ecosystem of decentralised applications (dApps) and blockchain-based services.

While Web3 is still not mainstream, it's rapidly gaining momentum. The promise of a more democratic and value-rich internet experience is resonating with many individuals and businesses alike. As more developers, entrepreneurs, and innovators enter the Web3 space, we can expect to see exciting new applications and use cases emerge.

Of course, with any new technology, there are challenges to overcome. Technical hurdles, scalability issues, and user adoption are all areas that require attention in the Web3 ecosystem. Interoperability, or the ability of different blockchain networks to communicate seamlessly, is a technical challenge that needs to be addressed. Standardisation efforts and innovative solutions are being developed to bridge these gaps and create a more interconnected Web3 ecosystem.

So, there we have it. Web3 represents a paradigm shift in the way we interact with the internet. It's not about replacing Web2 but rather complementing it with decentralisation and blockchain technology. Web3 adoption is still in the early stages, and we need to educate and market it to the masses. We need to show the early majority the benefits and potential pitfalls of Web3 in a relatable way. Scams and rug pulls are a bummer, but with education and community support, we can protect users and build trust. Improving the user experience is key, with easy interfaces, smooth onboarding, affordable fees, and tight security. Web3 ain't replacing Web2; it's working together, bringing new opportunities. It's a journey, but the future looks promising for a more democratic and value-driven digital world.

Chapter 12:
Crypto Cowboys and Tech Wizards: Navigating Web3 hiring

So, picture this: It's late 2022, and the Web3 space is feeling very negative after the LUNA and FTX debacles. I decided to take the plunge and start my own Web3 agency. But let me tell you, it was no walk in the park. We were smack in the middle of a bear market, and things were tough. However, I believed that the people who were still active in the space during that time were the real deal. They were the passionate ones, the ones who were in it for the long run. They didn't give up when things got hard. They had the experience and resilience to stick it out through the toughest times. Little did I know, I was about to encounter a whole new set of challenges when it came to hiring people in the Web3 space.

Let's start with the degens. If you're not familiar with the term, let me break it down for you: a Web3 degen is someone who loves to take high-stakes risks without fully understanding what they're getting into. They're the high-roller gamblers of the crypto world. These folks stay up all night trading and betting on short-term gains. They live for the adrenaline rush and the thrill of making quick profits. It's like a ride for them.

Now, there's nothing inherently wrong with being a degen. Some of them are really knowledgeable about crypto, the markets, NFTs, and all that jazz. They know their stuff when it comes to high-risk short-term trading. But here's the thing: hiring degens for a sustainable business is a whole different story.

You see, degens are used to fast results and quick dopamine hits. They're accustomed to looking at 1–5-minute charts, where price bars are opening and closing in the blink of an eye. They're in and out of trades within minutes, and sometimes they make a profit just as fast. It's an intense rush.

But building a sustainable business and team for the long term? That's a whole different ball game. It takes time, persistence, patience, and resilience. I learnt lessons from hiring degens in the early days of my business, and let me tell you, those particular degens didn't last long. I quickly learned that their impatience and short attention spans could negatively impact the rest of the team, our clients, and potential clients.

One thing I noticed about degens is that they tend to have a know-it-all attitude. They think they have all the answers when it comes to Web3, NFTs, and cryptocurrencies. They believe they know which projects will succeed and which ones will fail. And let me tell you, they're not afraid to voice their opinions, including to the point of openly criticising anyone they don't agree with in the Web3 space.

I've heard stories of degens actively seeking confrontation with Web3 company founders on Twitter Spaces, just to prove that they know more. But here's the kicker: when the project founders address their concerns and deal with them professionally, the degens often do a complete 180 and start promoting the very company they were criticising. It's a wild ride with these folks.

Now, I'm all for having a team that speaks their mind and pushes boundaries. Constructive criticism is important for growth. But when a degen tells a potential client that what they're building is rubbish without even taking the time to research, learn, and ask questions, well, that's a problem. It doesn't exactly create a good impression, does it?

Another thing I've noticed about degens is that they have a different sense of time. They're up until all hours of the morning, trading and connecting with their online network. Timetables and calendars are usually foreign concepts to them. So, it's not uncommon for a degen to show up late to a meeting or, in some cases, not show up at all. They're just not familiar with the professionalism, structure, and self-discipline that are required to operate effectively in a team and business environment.

Now, before you think that all degens are bad news, let me clarify that it's not the case. Some degens are incredibly knowledgeable and have valuable insights to offer. They can provide unique perspectives and contribute to the team in meaningful ways. However, it's important to proceed with caution and consider the potential challenges that may arise when working with degens.

On the flip side, we have the professionals. These are the individuals who have been in the Web3 space for a while. They understand the technology, the markets, industry dynamics, and bring a wealth of experience and a level-headed approach.

When you hire professionals, you can expect a more structured and disciplined approach. They understand the importance of long-term sustainability and building a solid foundation. They're not just in it for the quick gains, but for the growth and development of the industry as a whole.

One of the advantages of working with professionals is their expertise and deep knowledge of the Web3 ecosystem. They've done their homework, studied the projects, and can provide valuable insights and strategic guidance. They understand the nuances of different blockchain platforms, can analyse market trends, and help you make informed decisions.

Another benefit of working with professionals is their ability to maintain structure - they're familiar with timelines, deadlines, and the importance of being on time. You won't have to worry about them showing up late to meetings or disregarding the basic principles of professionalism.

Professionals also tend to have a wider network and connections within the Web3 space. This can be incredibly valuable when it comes to partnerships, collaborations, and accessing resources. They know the right people to reach out to and can help expand your opportunities.

However, it's important to note that not all professionals are created equal. Just like in any industry, you'll find a mix of skill levels and personalities. Some professionals may be more collaborative and open to new ideas, while others may be more set in their ways. It's important to find individuals who align with your values and vision.

When it comes to hiring in the Web3 space, it's all about finding the right balance. You want a team that is passionate, experienced, and committed for the long haul. Whether you choose to work with degens, professionals, or a mix of both, it's essential to have clear expectations, open communication, and a shared vision for the future.

Building a successful Web3 business requires assembling a team that can weather the storms, adapt to changes, and contribute their unique perspectives and expertise. It's an exciting journey, and with the right people by your side, you can navigate the Web3 landscape and make a meaningful impact in this rapidly evolving industry.

Chapter 13:
From Billboards to Blockchain: Marketing in Web3

As a marketer at heart, I couldn't write a book without a chapter about the topic of marketing.

First off, let's talk about the unique challenges and opportunities that come with marketing in the Web3 space. Traditional marketing tactics alone won't cut it here - it's a new ball game, my friend. Although in saying this, there are some core fundamentals regarding strategy and marketing that remain – and are important not to forget. We'll run through both.

One of the first things you'll notice is that the Web3 community is incredibly tech-savvy and well-informed. They're not easily swayed by flashy ads or empty promises. They want to see real value and authenticity.

When it comes to marketing in Web3, trust is everything. The native Web3 community is all about breaking free from the clutches of traditional advertising and taking control of their own lives. They're not interested in being pawns in some big marketing game. Nope, they want to know who you really are, what you believe in, and how you're making a positive impact. So, the key here is transparency. Be open, honest, and genuine in your communication. Engage in real conversations with the community and show them that you truly care about their needs and aspirations.

Web3 enthusiasts want to see real value and authenticity. So, drop the corporate jargon and speak their language. Show them that you're one of them, that you understand their desires and concerns. Connect with them on a personal level and build that trust brick by brick.

One way to build trust is by being transparent about your intentions and actions. Let them know what you stand for and how you're making a difference. Share your values, your goals, and your vision for the future. Don't be afraid to admit your mistakes or address any concerns that may arise. The Web3 community appreciates honesty and vulnerability.

But trust isn't just built through words. It's also about actions. Show the community that you're not just in it for the quick buck. Demonstrate your commitment to their well-being and the greater good. Support causes that align with their values, engage in sustainable practices, and give back to the community. Actions speak louder than words, as the old saying goes.

Engaging in open dialogue is another crucial aspect of Web3 marketing. This community thrives on collaboration and participation. They want to be heard and have their opinions valued. So, create spaces where they can voice their thoughts, concerns, and ideas. Actively listen to their feedback and respond genuinely. When you involve the community in your decision-making process, you're not just building trust, but also fostering a sense of ownership and belonging.

Remember, in Web3, it's not just about selling a product or service - it's about building a community and creating a movement. So, be a part of that movement. Be transparent, engage in open dialogue, and demonstrate through your actions that you genuinely care. That's how you build trust in the wild world of Web3 marketing.

Alright, let's talk about the importance of education in Web3 marketing. This stuff can get pretty confusing, I won't lie. Blockchain, NFTs, decentralised this, and decentralised that - it's like a whole new language!

You see, the Web3 space is still pretty fresh and shiny. A lot of folks are still scratching their heads, trying to figure out what it's all about. Taking those complex concepts and turning them into something that people can actually wrap their minds around is what it is all about. In my experience of educating our team and then our Web2 clients, using relatable examples, telling stories, and painting a picture of how Web3 can actually make a difference in their lives is a winner.

One of the key things you need to do is show people Web3's benefits and possibilities and help them understand how this new technology can enhance their day-to-day existence. Maybe it's about owning their data and having control over their digital identity. Or perhaps it's about unlocking new economic opportunities and empowering individuals to thrive in a decentralised world. Or maybe it's giving them a real example of how they can implement this technology in their business to improve their current situation. Whatever it is, make it tangible and relatable.

Now, let's talk about empowerment. Web3 is all about giving power back to the people and breaking free from those pesky big corporations. Gone are the days of being passive consumers. In the realm of Web3, we have the power to be active participants and co-creators and Web3 is the ticket to being part of something bigger and having a real impact.

But here's the thing – not everyone understands the true potential of Web3. As Web3 natives, it's our mission is to help people realise that Web3 is their ticket to empowerment. Web3 gives us the tools and the platform to be the architects of our own destiny.

So, how do we convey this sense of empowerment? By showing people real-life examples of individuals and communities who have embraced Web3 and are making waves. There will always be naysayers who think Web3 is just a passing fad. We need to help people see that it's not just a trend – it's a movement that's here to stay.

As I was finalising this book in late 2023, I attended an event in London where Steven Bartlett was a headline speaker. What he said about web3 resonated so well. He said web3/crypto has so much backlash and people fighting against it because something big and meaningful is at stake. And in history, all technology advancements have come with push back, and especially when something important is at stake. And that's what makes him believe in the future of Web3 even more.

Now, let's talk about community-building. Here's the deal: in this realm, communities are everything. They're like the heartbeat of the whole ecosystem. Marketing's mission is to nurture and grow a thriving community around the brand or project. It's all about fostering a sense of belonging, collaboration, and providing value.

Think of it this way: we want people to feel like they're part of a bigger family. We're not just selling a product or service; we're building a community where folks can come together, share ideas, and support one another. It's like having a bunch of like-minded friends who are all passionate about the same thing.

We do this by creating spaces where people can connect, exchange ideas, and learn from one another. It could be a forum, a Discord server, or even regular virtual meetups. The goal is to provide a platform for meaningful interactions and knowledge-sharing.

Influencer marketing also plays a significant role in Web3. But here's the thing: Web3 influencers are not your typical influencers. They're not just after a quick pay cheque. They genuinely believe in the projects they promote and have a deep understanding of the technology behind them. So, when you're looking for influencers to collaborate with, make sure they align with your values and have the knowledge to back it up.

Another aspect of Web3 marketing is embracing the power of user-generated content. Web3 enthusiasts are a creative bunch, and they love to get involved. Encourage your community to create and share content related to your brand or project. It could be anything from artwork, memes, videos, or even blog posts. User-generated content not only builds engagement but also helps spread the word organically.

Data collection is another important piece of Web3 marketing. We've spoken about how big tech giants need to change their ways, and that's where Web3 has come into play. In the Web3 world, we need to be more considerate of data collection and how it is used. Airdrops are a particularly interesting way of overcoming the data collection issue in Web3.

Let's go deeper with airdrops and how they can be a game-changer in Web3 marketing. Picture this: you're strolling through the Web3 landscape, and suddenly, out of the blue, you stumble upon a treasure trove of free tokens raining down from the sky. That's what an airdrop feels like - a gift from the crypto gods themselves.

So, what exactly is an airdrop? Well, it's when a project or platform decides to distribute something for free to a group of people. It's like a digital version of throwing goodies into a crowd at a concert, except in this case, the goodies can be tokens with real value. It's a way for projects to get their name out there, build awareness, and reward early adopters.

Now, you might be wondering how airdrops can be used for Web3 marketing. Airdrops are like marketing magic because they create a buzz and generate excitement. When people hear that free tokens are up for grabs, they can't help but flock to the project or platform. It's like a digital feeding frenzy!

But here's the key: airdrops are not just about giving away tokens. They're about building a community and fostering engagement. When people receive these tokens, they become stakeholders in the project. They have a vested interest in its success. And that's where the marketing magic comes in. These newfound token holders become brand advocates. They start talking about the project, sharing it with their friends, and spreading the word like wildfire.

Think about it - if someone gave you free tokens, wouldn't you be excited to see them grow in value? You'd probably tell your buddies about it, right? And that's how airdrops can create a ripple effect in the Web3 world. It's like planting seeds of excitement and watching them grow into a flourishing community.

But airdrops aren't just about attracting new users. They're also a way to reward and retain existing community members.

It's a gesture of appreciation for their support and loyalty. By distributing tokens to those who have been with you from the beginning, you're showing them they're an integral part of your journey.

Now, I should mention that airdrops aren't a one-size-fits-all solution. They need to be carefully planned and executed. You want to target the right audience – those who are genuinely interested in your project or platform. You need to find the people who align with your values and have the potential to become long-term supporters.

So, as you can see there are differences and challenges when it comes to Web3 marketing. However, I mentioned that there are core marketing and strategy fundamentals that should not be lost just because we're moving into a Web3 world. Let's talk about these core marketing and strategy fundamentals that remain rock-solid even in the wild world of Web3. While Web3 might be all about decentralisation, blockchain, and fancy crypto stuff, some things never change when it comes to marketing.

First and foremost, you must know your audience. Knowing your audience is marketing 101. It's like trying to hit a bullseye in darts – you need to know who you're aiming for. Understand their needs, desires, and pain points. Figure out what makes them tick and how your product or service can make their lives better. It's important to map out your target market and even define the customer personas of those you want to reach. This will help you determine which platforms they spend most of their time on, so you can be confident that you're not wasting your time on other platforms.

Next up, we can't forget about storytelling. Humans have been telling stories since the dawn of time, and it's still the most powerful way to connect with people. So, in the Web3 world, it's all about crafting a compelling narrative that resonates with your audience. Tell them why your project matters, how it can solve their problems, and why they should care. Engage their emotions and make them feel like they're part of something bigger than themselves.

Another fundamental principle is building trust - trust is the glue that holds everything together. People won't just hand over their hard-earned crypto or invest in your company if they don't trust you. So, be transparent, honest, and deliver on your promises. Show them that you're here for the long haul and that you genuinely care about their well-being. Trust is like a delicate plant – nurture it, and it will grow strong.

Let's not forget about differentiation. In the Web3 space, competition can be fierce. There's a lot of noise and shiny projects vying for attention. So, you need to find your unique selling proposition – that thing that sets you apart from the crowd. Maybe it's a ground-breaking technology, a revolutionary concept, or simply a better user experience. Whatever it is, make sure you stand out in the sea of Web3 activations.

Now, let's talk about metrics. Numbers are crucial to measure your success. Whether it's Web3 or not, you need to track key performance indicators (KPIs) to see if your marketing efforts are paying off. Keep an eye on metrics like user engagement, conversion rates, and customer acquisition costs. This way, you can fine-tune your strategy, pivot if needed, and make data-driven decisions.

Last but not least, partnerships are still a powerful tool in Web3 marketing. Just like in the traditional marketing world, collaborating with other projects or influencers can amplify your reach and credibility. Find partners who share your values and have a similar target audience. By joining forces, you can create a win-win situation and tap into new markets.

While Web3 marketing may seem like uncharted territory, the core marketing and strategy fundamentals remain as solid as ever. Know your audience, tell compelling stories, build trust, differentiate yourself, track your metrics, and seek out strategic partnerships. These timeless principles will guide you through the Web3 maze and help you stand out.

Chapter 14:
Breaking Free from the Apes and Vapes Mindset

I f you've been exploring Web3, you may have come across a peculiar phenomenon - a seemingly overwhelming obsession with cartoon ape NFTs, festivals/parties and an unexpected association with vaping. From the outside looking in, it's easy to assume that Web3 is all about apes, vapes, and parties. However, after reading this book and learning about my journey and passion for Web3, you will now understand that there is so much more to it!

At its heart, Web3 is all about decentralisation and distributing power and decision-making authority across a network of peers. This decentralised model ensures that no single entity has absolute control, reducing the risks of censorship and manipulation. In Web3, the power truly lies with the people, fostering trust and empowering individuals. One of the most exciting aspects of Web3 is the emphasis on data ownership and putting individuals back in control of their digital footprints. Web3 technologies mean individuals can securely store their data on decentralised networks like the blockchain. They have the ability to decide who can access their data and under what circumstances.

Web3 represents a paradigm shift, challenging the dominance of global corporate giants that have historically controlled our online experiences.

Now, let's address the perception that Web3 is all about apes, vapes, and parties. While it's true that these elements have gained attention, they merely represent one aspect of the diverse and evolving Web3 community. The popularity of cartoon ape NFTs and vaping culture is more a reflection of the vibrant and creative spirit within the community than a definition of Web3 itself.

Web3 encompasses a wide range of projects, initiatives, and technologies. It's about the development of decentralised finance (DeFi), enabling individuals to access financial services without relying on intermediaries. It's about the creation of social platforms that prioritise privacy, security, and user control. It's about open-source collaborations, fostering innovation and collective problem-solving. Web3 is a playground for entrepreneurs, developers, artists, thinkers, and enthusiasts, all united by a shared vision of a more inclusive and user-centric internet.

Web3 is much more than a fleeting obsession with cartoon apes and vaping. So, let's look beyond the surface and together shape a future that puts power back into the hands of the people.

Conclusion:
Web3 is our Take 3

Web3 is more than just the latest buzzword or trend—it's our take 3, our third chance to get things right in the digital realm.

First things first, let's recap a bit. Web1 was the pioneer, the OG of the internet. It brought us together, connected us across borders, and opened up a world of possibilities. But it had its flaws, right? It was centralised, controlled by a few big players who dictated the rules. We had limited control over our data, and trust issues arose as we witnessed breaches and misuse.

Along came Web2, the era of social media, smartphones, and apps galore. It was a wild ride, full of exciting innovations and instant connectivity. We could share our lives, connect with friends and family, and access information like never before. But let's be real—it also brought its fair share of headaches. Our personal data became a commodity, sold to the highest bidder. We got trapped in filter bubbles, echo chambers, and the constant barrage of clickbait. The power was concentrated in the hands of a few corporate giants who called the shots.

Enter Web3, our take 3. It's our third chance to reshape the digital landscape, to learn from the past and create a better future. Instead of relying on a central authority, Web3 distributes power among a network of peers. It's like a digital democracy, where everyone has a say. This paves the way for more transparency, accountability, and fairness. We're taking back control, my friends!

With Web3, we're not just users; we're active participants. We have a stake in the game. Through blockchain technology, we can engage in peer-to-peer interactions, eliminating the need for middlemen. We can transact, create, and collaborate directly with others. Say goodbye to the days of relying on a faceless corporation to approve our ideas or validate our worth.

With Web3, we're reclaiming our digital sovereignty.

So, let's remember that Web3 is our take 3. It's our opportunity to learn from the past, to correct the mistakes, and to shape the digital realm in a way that benefits us all. It's time to dream big, take risks, and build a digital future that's more fair, transparent, and inclusive. Here's to our take 3!

Acknowledgments

First and foremost, I want to give a huge shout-out to my partner of more than 16 years, Damien. I wouldn't be where I am today without your unconditional support. From the bottom of my heart, thank you for being there for me.

Damien, you've been more than just a partner; you've been my biggest believer. When I was unsure of myself, you were there, cheering me on and reminding me of my unlimited potential. You've been my rock, constantly encouraging me to chase my dreams and assuring me that I can achieve anything I aim for.

So, Damien, thank you. Thank you for being my partner for 16+ years, my best friend, and my number one fan. Thank you for believing in me, even when I doubted myself. Thank you for the love, the encouragement, and the unwavering support that has propelled me forward. I am forever grateful to have you in my life.

I also want to give a special shout-out to my mentor and coach, Arnon. Arnon, you've been instrumental in my journey and the reason I'm writing this book. Your encouragement, support, and wise counsel have been invaluable to me. I can't thank you enough for pushing me beyond my limits.

To my incredible team, you guys rock! Through thick and thin, you've stood by my side, weathering the storm with me. Your dedication, hard work, and unwavering belief in our vision have been nothing short of inspiring. I couldn't ask for a more talented and committed group of people to work with.

I'd also like to thank Marcus de Maria and the team at Investment Mastery; Dennis, Eamon and Keith in particular. You opened my eyes to this new world (and answered my many questions along the way!). Thank you for the education, encouragement and support. Your patience and generosity is something I will always be grateful for.

To Sandy, thank you. I refrained from talking about you too much in this book, as I understand you value your privacy. It really all started with you, though! Thanks for showing me the light. Your teachings have been invaluable.

And how can I forget my Web3 network? Each and every one of you has left an indelible mark on me, inspiring me in your unique ways. The passion, creativity, and innovative spirit that I've witnessed within this community are nothing short of mind-blowing. Your drive to challenge the status quo, reimagine possibilities, and shape a better future has fuelled my own fire. I am eternally grateful for the inspiration and camaraderie we share.

To all those who have supported me along this wild ride, whether it's been through words of encouragement, lending an ear during tough times, or simply being there to celebrate the wins – thank you. Your belief in me and what I'm trying to accomplish means the world.

Each person mentioned here has played a vital role in shaping my path and journey, and for that I say, thank you.

www.ingramcontent.com/pod-product-compliance
Lightning Source LLC
Chambersburg PA
CBHW040757220326
41597CB00029BB/4975